TRB Special Report 197

Transportation of Hazardous Materials: Toward a National Strategy (Volume 1)

in cooperation with
U.S. Department of Transportation
Research and Special Projects Administration
Federal Emergency Management Agency
National Association of Governors Highway Safety Representatives

Transportation Research Board
National Academy of Sciences
National Research Council
Washington, D.C. 1983

Transportation Research Board Special Report 197 (Volume 1)
Price $8.00
Edited for TRB by Brenda J. Vumbaco

modes
all

subject area
51 transportation safety

Transportation Research Board publications are available by ordering directly from TRB. They may also be obtained on a regular basis through organizational or individual affiliation with TRB; affiliates or library subscribers are eligible for substantial discounts. For further information, write to the Transportation Research Board, National Academy of Sciences, 2101 Constitution Avenue, N.W., Washington, DC 20418.

Notice
The project that is the subject of this report was approved by the Governing Board of the National Research Council, whose members are drawn from the councils of the National Academy of Sciences, the National Academy of Engineering, and the Institute of Medicine. The members of the committee responsible for the report were chosen for their special competence and with regard for appropriate balance.

This report has been reviewed by a group other than the authors according to procedures approved by a Report Review Committee consisting of members of the National Academy of Sciences, the National Academy of Engineering, and the Institute of Medicine.

The views expressed in this report are those of the authors and do not necessarily reflect the view of the committee, the Transportation Research Board, the National Academy of Sciences, or the sponsors of the project.

Library of Congress Cataloging in Publication Data
National Research Council. Transportation Research Board.
Transportation of hazardous materials.

(TRB special report; 197)
Proceedings of a conference held in Williamsburg, Va., Feb. 17-20, 1981.
1. Hazardous substances–United States–Transportation–Congresses. I. National Research Council (U.S.). Transportation Research Board. II. Series: Special report (National Research Council (U.S.). Transportation Research Board); 197.
HE199.5.D3T72 1983 363.1'7 83-5002
ISBN 0-309-03510-4 (set) ISSN 0360-859X

Sponsorship of the Papers in This Transportation Research Board Special Report

Steering Committee to Develop a National Strategy for the Transportation of Hazardous Materials and Hazardous Wastes in the 1980s

Chairman

Raymond D. Scanlon (October 1982-present)
Karsten J. Vieg (1980-1982)

Members

Lawrence W. Bierlein, William A. Brobst, Gregory R. Choppin, Robert M. Graziano, Erskine E. Harton, Jr., Jerry A. Havens, Robert M. Jefferson, William E. Johns, Jack T. Lamkin, Jeremiah J. Driscoll, Dennis L. Price, Deborah Karen Schultz Rudolph, Eugene R. Russell, George L. Wilson, Jr. (deceased)

Transportation Research Board Staff

Adrian G. Clary

Conference Consultant

Dennis L. Price

The organizational units, officers, and members are as of December 31, 1982.

Contents

ACKNOWLEDGMENT

This report was prepared by Dennis L. Price, Virginia Polytechnic Institute and State University, Blacksburg, and by Kenneth B. Johns, Adrian G. Clary, and Janice W. Bain of the Transportation Research Board (TRB) staff, who borrowed freely and often verbatim from both workshop and resource materials generated by the National Strategies Conference on the Transportation of Hazardous Materials and Wastes in the 1980s. The conference, which took place February 17-20, 1981, in Williamsburg, Virginia, was conducted by TRB.

The titles of the resource papers and their authors are as follows:

Title	Author
Regulation: Part 1. Purpose and Direction of Regulation	C. H. Thompson
Regulation: Part 2. The Process of Hazardous Materials Regulation	L. W. Bierlein
Regulation: Part 3. Complexity of Hazardous Materials Transportation Regulations	D. A. Boyd
A Question of Training	A. C. Bensmiller
Segment I: Training Concepts Assessed	J. Granito
Segment II: Technical Training	F. Halvorsen
Emergency Response	J. J. O'Driscoll B. L. Hansen R. J. Mesler, Jr.
Civil Liability and Social Regulation	S. Hoffman
Criminal Sanctions and Regulating Corporate Behavior in Transportation of Hazardous Materials	H. A. Johnson
Preemption: How Do We Deal with Interjurisdictional Conflicts with Law?	J. K. Healy
Hazardous Materials Transportation Risk Assessment	L. L. Philipson H. S. Napadensky M. N. Maxey
How-to-Do-It Regulations Inhibit Research	W. C. Jennings
Government Role in Fostering Innovation	S. Prensky
Application of Automated Data Base Technology to an Intense Regulatory Climate	D. M. Shilesky

The discussion leaders and rapporteurs for the six workshops are noted below:

Workshop	Discussion Leader	Rapporteur
Regulation	C. H. Thompson	D. Rudolph
Training	A. C. Bensmiller	G. R. Choppin
Emergency Response	R. J. Mesler	R. Graziano
Legal Responsibilities and Implications	H. A. Johnson	S. M. Wasser
Risk Assessment	L. L. Philipson	T. S. Glickman
Technological Innovation Needs and Limits	J. Havens	W. A. Brobst

In addition, special appreciation is extended to Steering Committee members Lawrence W. Bierlein, Karsten J. Vieg (Chairman, 1980-October 1982), and Raymond D. Scanlon (Chairman, October 1982-present) for their assistance in writing, reviewing, and editing this report. Special appreciation is also extended to Peter Sill, Research and Special Programs Administration, U.S. Department of Transportation; James Kerr, Federal Emergency Management Agency; and Joe Murphy, National Association of Governors Highway Safety Representatives. Their financial support at critical points made this project possible.

The Rail Vehicle Safety Research Division of the Federal Railroad Administration, together with the National Transportation Safety Board, made a very valuable contribution in the form of a legislative history of hazardous materials transportation, which was distributed to all conference participants. The compendium, <u>Federal Legislation Affecting the Transportation of Hazardous Materials, With Selected Bibliographies and Data</u>, will be available from the National Technical Information Service in early 1983.

PART 1
Introduction

TRB ROLE

BACKGROUND

Although the report that follows had an immediate germination period of approximately two-and-a-half years, in a larger sense it is among the most recent efforts in a 117-year process to ensure safety in the transportation of hazardous materials and hazardous wastes in the United States. Precedent within the Transportation Research Board (TRB) for convening and analyzing the best national thinking on this issue was set by the 1969 Conference on Hazardous Materials at Airlie House, Warrenton, Virginia. That conference, the result of a request by the U.S. Department of Transportation (DOT) to the National Research Council, assembled 90 people from government and industry to review the situation as it then stood.

The Airlie House recommendations (17) paved the way for some of the reforms that were implemented during the 1970s. Legislatively, the most significant was the passage of the Hazardous Materials Transportation Act (HMTA) of 1975. Administratively, the Materials Transportation Bureau (MTB) in DOT's Research and Special Projects Administration (RSPA) was established. TRB's Standing Technical Committee on the Transportation of Hazardous Materials also grew out of the Airlie House deliberations.

COMMITTEE ACTIVITIES

Prompted by growing public and congressional concern over hazardous materials transportation in the late 1970s--a concern that resulted from a number of well-publicized hazardous materials incidents and spills (some involving multiple fatalities)--TRB's Committee on the Transportation of Hazardous Materials, in developing program plans, initiated a number of activities designed to further its role as noted below in the Committee's scope:

> This committee is concerned with the movement of hazardous materials including conditions and forces encountered during transportation, type and extent of hazards associated with material type of class, preservation and packaging to prevent cargos' damage or decomposition during transportation and handling, laws controlling and legal liabilities pertaining to hazardous materials movements, enforcement and administrative controls and procedures for such movements, signs and labels to indicate the hazardous nature of the cargo, and the knowledge of incidents occurring during transportation and handling movement risks.

Specifically, the Commiteee undertook, and published the results of, a survey to identify the ten most critical issues in hazardous materials transportation; initiated the 1981 Williamsburg Conference; and organized TRB Annual Meeting sessions that have covered all aspects of hazardous materials transportation and from which many of the papers have been published and distributed widely as part of TRB's information dissemination role.

IDENTIFICATION OF CRITICAL ISSUES

In 1980, TRB's Committee on the Transportation of Hazardous Materials issued a list of the ten most critical issues in the transportation of hazardous materials (37). The issues were derived from responses to a questionnaire that contained a list of 25 issues deemed by the Committee to be critical in nature. The responses came from federal, state, and local governments; industry; and trade and professional associations. Respondents were asked to identify the ten most critical issues and to add others to the list that they felt were important. The final listing was published under the auspices of the Committee as a basis for a reasoned approach and a common strategy to improve the nation's hazardous materials transportation system. The issues, not listed in order of importance, fall into five categories: regulations (issues 1 and 2); need for a coordinated systems approach (issues 3-6); data and data applications (issues 7 and 8); legal responsibilities (issue 9); and public awareness (issue 10). The ten most critical issues listed are as follows:

1. [Need for] Harmonious International, Federal, State, and Local Hazardous Matrerials Regulatory Controls;
2. The Complexity of DOT's Hazardous Materials Regulations and the Need to Convert Some of Them from Detailed Specifications to Performance-Based Criteria;
3. [Need for] a National Strategy for Control of Hazardous Materials Risks;
4. Training for All Persons Involved in the Transportation of Hazardous Materials, Including Shippers, Carriers, and Emergency-Response Personnel;
5. [Need for] a Single, National Response System for Incidents and Accidents Involving the Transportation of Hazardous Materials;
6. [Need for] an Integrated Hazardous Materials Transportation Administrative Communication System Among Federal and State Governments;
7. [Lack of] a Comprehensive Data System for the Flow of Hazardous Materials by Quantity, General Hazard Class, Route, and Mode;
8. [Need to Synthesize and Evaluate for Its Practicality] the State of the Art for Hazardous Materials Transportation Cost-Benefit-Risk Analysis Methodology;
9. Clarification of the Legal Responsibilities of Governmental and Private Agencies Involved in Hazardous Materials Transportation; and
10. [Need for Public Awareness] About the Relative Safety of Hazardous Materials Transportation.

These issues were subsequently telescoped into six discussion topics for the 1981 Williamsburg conference.

NATIONAL STRATEGIES CONFERENCE

A Steering Committee to Develop a National Strategy for the Transportation of Hazardous Materials and Hazardous Wastes in the 1980s was appointed, with the approval of the Chairman of the National Research Council, in October 1980. The Steering Committee was composed of selected members of TRB's Standing Technical Committee as well as members who represented government agencies, industry, and academia, and whose expertise covered some facet of the handling, shipping, regulation, enforcement, safety, or legal aspects of hazardous materials transportation. The Committee was chaired by Karsten J. Vieg, then Director of Traffic Safety, Illinois Department of Transportation.

The purpose of the conference, held under the Steering Committee's aegis, was to develop recommendations for a comprehensive national strategy to provide safe and efficient transportation of hazardous materials and hazardous wastes in the 1980s. The conference covered the following aspects of the hazardous materials transportation system: regulation, training, emergency response, legal responsibilities and implications, technological needs and limitations, and risk assessment. Consideration of the public interest was a pre-eminent theme underlying all the discussions.

The conference, held February 17-20, 1981, in Williamsburg, Virginia, was funded in part by the U.S. Department of Transportation's Research and Special Projects Administration; the Federal Emergency Management Agency; and the National Association of Governors Highway Safety Representatives. Attendance at the conference was by invitation--extended to hazardous materials transportation experts and policymakers who represented the executive and legislative branches of the federal government; state and local government agencies; and the private sector.

The 150 conference participants were pre-assigned to two of the six topical discussion groups and received copies of the appropriate resource papers. Background papers for each of the six discussion topics were presented on the first day of the conference, after which participants attended two sessions of each of two workshops. Rapporteurs' reports, presented on the last day of the conference, highlighted issues/problems attendant to each of the discussion topics and made appropriate recommendations.

The six topics derived from the critical issues were the basis for 14 resource papers commissioned by the Steering Committee as a catalyst for discussion. These resource papers, together with the workshop rapporteurs' reports, provide a frame of reference for the Steering Committee's findings and recommendations, which are discussed in subsequent sections of the report.

CONFERENCE REPORT

The findings and recommendations of the Steering Committee, developed pursuant to the Williamsburg meeting, are based on the findings and recommendations of the six discussion groups; conferencewide discussions; the resource papers; and the collective judgment of the Steering Committee members and experts in the field of hazardous materials transportation. The findings and recommendations are summarized in Part 2 of this report; they are then discussed within the context of the broader issues they represent in Part 4. Part 3 of the report is an overview of the background history, legislation, regulations, and current issues inherent in the field of hazardous materials transportation and identifies the government and private agencies that play major roles in the field. Part 5 of the report is composed of three Appendixes: references, a list of conference attendees, and a roster of the Steering Committee.

The report deliberately does not give lengthy treatment to marine or air transport of hazardous materials. This is because much of the marine traffic is international and is governed by international regulations that are beyond the scope of national strategies. Air traffic in hazardous materials is minimal, and the work of international and national trade and professional organizations has been effective in relieving many problems in this area.

The transportation of nuclear wastes and radioactive materials was not a focus of attention at the conference, largely because it is a relatively small part of current hazardous materials transportation. Furthermore, coupled as it is with public concerns about the disposal of radioactive wastes and disposal area sites, it is a volatile issue that could inappropriately mask many other important issues. However, the increasing number of nuclear materials and wastes being transported indicates that this issue is likely to command greater attention at future conferences.

This report constitutes a summary of the conference proceedings and the Steering Committee's subsequent deliberations in developing its findings and recommendations regarding the development of a national strategy for the movement of hazardous materials and wastes. A companion volume (Special Report 197, Volume 2) containing the conference resource papers and the rapporteurs' workshop summaries is available for readers who wish more detail or background on the six topics discussed.

In addition to the Airlie House report (17), this is the second report to deal with hazardous materials transportation issues emanating from the National Research Council in recent years. The other report, prepared by the Asembly of Engineering's Committee on Transportation, is titled A Review of the Department of Transportation Research and Special Programs Administration's Hazardous Materials Research and Development Program (16) and was prepared for the U.S. Department of Transportation in 1980. A report dealing with methodological approaches for risk assessment of hazardous materials transportation at state and local levels will be published in 1983 in the National Cooperative Highway Research Program (NCHRP) Synthesis of Highway Practice series. NCHRP is administered by TRB's Division D, Cooperative Research Programs.

PART 2
Summary of Committee Findings and Recommendations

Findings and Recommendations

The findings and recommendations that were developed by the TRB Steering Committee on the Transportation of Hazardous Materials following the National Strategies Conference on the Transportation of Hazardous Materials and Hazardous Wastes in the 1980s are summarized here. They are discussed in detail in Part 4 of this report.

FINDING 1

There should be a single, lead federal agency responsible and accountable for the safe transportation of hazardous materials.

RECOMMENDATION 1

The U.S. Department of Transportation (DOT) should properly be the lead federal agency for the development of transportation policy and procedures for hazardous materials.

FINDING 2

There is no agreed-on hazardous materials transportation safety goal within the government-industry complex.

RECOMMENDATION 2

A statement of purpose for the national hazardous materials transportation regulation program should be formulated and adopted.

FINDING 3

Related to Finding 2 is the fact that the relative roles of federal, state, and local governments have never been clearly and understandably defined, which has led to interjurisdictional confusion. In addition, no developed concept exists concerning how best to achieve compliance with those measures that will provide the greatest safety.

RECOMMENDATION 3

DOT should engage in a vigorous program to obtain a clear definition of roles for local, state, and federal officials engaged in regulation, enforcement of regulations, and response to incidents in hazardous materials transportation. In addition, state and local governments should have distinct roles in attaining compliance through the enforcement of hazardous materials regulations.

FINDING 4

Given that no system of regulation and enforcement can eliminate accidents, a reasoned response to such occurrences is important. The Steering Committee found that there is considerable disagreement on the procedures and adequacy of emergency responses. However, there is agreement that accurate, timely, and continuous information flow is necessary at the scene of the accident in order to properly evaluate the hazard involved.

RECOMMENDATION 4

Federal agencies with emergency-response duties should take action to develop required information and technology to provide accurate and fast identification and estimation (both remote and on scene) of the nature and degree of hazard in accidents and spills. The Congress should support research and development to improve evaluation of on-scene hazards and to facilitate communications between all responsible individuals and teams as well as with the general public.

FINDING 5

There is no systematic federal approach to training in the area of hazardous materials transportation.

RECOMMENDATION 5

DOT should designate an office to develop a master program of hazardous materials training for personnel in public and private organizations and for all phases of the transportation of hazardous materials.

FINDING 6

The Steering Committee identified a need for the inclusion of more technological innovation, information, and research related to the transportation of hazardous materials.

RECOMMENDATION 6

DOT should continue to address and press for implementation of technological developments that would improve hazardous materials transportation.

FINDING 7

Although encouraged by DOT's adoption of many of the recommendations from the 1969 Airlie House Conference, the Steering Committee also noted that some of the most far-reaching proposals made there, e.g., performance standards, have not been adopted.

RECOMMENDATION 7

Regulations should be improved by incorporating performance standards where it is feasible to do so. [Ed. note: Docket HM-181: Performance-Oriented Packaging Standards, announced April 15, 1982, is a move in this direction.]

FINDING 8

The lack of safety in the transportation of hazardous materials may often be exaggerated to the general public.

RECOMMENDATION 8

There should be a concerted government-industry effort to communicate to the public the relative safety that does exist in the transportation of hazardous materials and hazardous wastes.

PART 3

Overview of the Problem

Overview

The term "hazardous materials" is defined both by statute (the Hazardous Materials Transportation Act) and by regulation (49 CFR 171.8) to be those materials or substances in a form or quantity that the Secretary of Transportation has found to pose unreasonable risks to health and safety or property in commerce. The approximately 2400 materials so designated are listed in the Hazardous Materials Table (49 CFR 172.101 and 102) according to proper shipping name and description. These materials may be further classified as explosives, flammables, corrosives, combustibles, and poisons. Radioactive materials and etiologic (human-disease-causing) agents are also included in the definition.

Some of the more common materials included in these categories are anhydrous ammonia, chlorine, methane, ethylene [liquefied petroleum gas (LPG), propane], nitric acid, hydrogen cyanide, hydrocarbon fuels, and sulfuric acid. These materials, together with many others, are vital to the economy of the United States and are shipped, often in large quantities, by several modes. Because they are hazardous materials, there is an inherent risk in their transport.

TRAFFIC

There are insufficient data from which to derive verifiable estimates of the movement of hazardous materials. The most commonly reported estimate by DOT is that there are about 4 billion tons of hazardous materials shipped each year (24). There is, however, no documentation to support this figure, and neither government nor industry is able at present to be more specific, except to indicate that the hazardous materials industry is growing, as is the generation of hazardous wastes. The U.S. Environmental Protection Agency (EPA) estimates that on the order of 56 million metric tons of hazardous wastes are being produced each year, with a 4-6 percent increase annually (43).

Movement of hazardous materials by mode has not been quantified, except in the case of a few state surveys. Although some general figures as repeatedly reported in the literature give some idea of the size and scope of the traffic, the discrepancies in the figures also serve to further highlight the problem of the lack of reliable data.

Highways

DOT estimates that 5-15 percent of all trucks on the road at any given time carry hazardous materials (24). The American Automobile Association estimates that 2 billion tons of hazardous materials are transported by highway vehicles annually (30). The National Transportation Safety Board (NTSB) indicates that truck transportation accounts for about 30 percent of all hazardous materials transportation (24). DOT estimates that there are about 10,700 shippers and 11,700 carriers involved in the highway movement of hazardous materials. There are thought to be at least 413,000 tank trucks regularly transporting hazardous materials (24).

Railroads

The Association of American Railroads estimates that more than 1 million carloads (80 million tons) of hazardous materials, primarily chemicals, are transported annually, largely by tank cars that constitute 10 percent of the U.S. rail car fleet (30). This tonnage is thought to represent about 70 percent of the total hazardous materials movement by all modes. Chemical traffic, of which 75 percent is considered hazardous, is increasing--23 percent from 1975 through 1979 (15).

TYPES OF MATERIALS TRANSPORTED

In a 1979 study for the U.S. Senate, the Congressional Research Service listed hazardous substances in order of frequency of transport as follows (4):

Substance	Percentage of Shipments
Gasoline and jet fuels	56
Distillate fuel oil	34
Anhydrous ammonia	4
LPG	2
Paints and allied products	2
Industrial gases (compressed and liquefied)	1
Other	1
	100

Of all hazardous substances, those implicated in the highest percentage of incidents include the following (4):

Substance	Percentage
Paint and related products	37
Batteries (electric storage, wet)	11
Gasoline	9
Compounds (cleaning, liquid corrosive)	8
Compounds (paint removing, flammable liquid)	7
Other	28
	100

In the highway mode, the list changes. Gasoline accounts for 17 percent of the incidents; sulfuric acid, 11 percent; anhydrous ammonia, 5 percent; and caustic soda, hydrochloric acid, and LPG, 4 percent, respectively (4).

In rail transport, sulfuric acid, phosphoric acid, anhydrous ammonia, hydrochloric acid, corrosive liquids, and flammable liquids are in that order the major chemicals involved in rail incidents (31).

SAFETY OF TRANSPORT

Carrier Incidents

Establishing trends in accidents in hazardous materials transportation is difficult. Prior to 1971, no centralized, standardized system existed to provide uniform data for such accidents. When such a system was put in place by DOT, reported incidents rose from slightly more than 2000 in 1971 to more than 17,500 in 1979. It is believed that the enforcement of reporting requirements for the data system is more responsible for this almost ninefold increase than an actual increase in incidents because the number of deaths and injuries remained at a nearly constant level during the period.

Reportable incidents are those transportation events (including those occurring during loading, unloading, and temporary storage) in which, as a direct result of hazardous materials, (a) a person is killed; (b) a person receives injuries that require hospitalization; (c) estimated carrier or other property damage exceeds $50,000; (d) fire, breakage, spillage, or suspected radioactive contamination occurs involving shipment of radioactive material; (e) fire, breakage, or suspected contamination occurs involving shipment of etiologic agents; or (f) a situation exists of such a nature that, in the judgment of the carrier, it should be reported (49 CFR 171.15). These require immediate notification--that is, notification at the earliest practical moment--and a detailed report must be given within 15 days of discovery. Furthermore, each incident that occurs during the course of transportation in which there has been an unintentional release of hazardous materials from a package or tank must be reported in detail within 15 days of discovery (49 CFR 171.16). While the total number of incidents reported yearly may seem large, many do not involve serious consequences. For example, of the 16,115 reported incidents in 1980, only 324 (2.0 percent) were "severe"--i.e., causing fatality, injury, or requiring that the surrounding area be evacuated (41). In the subsequent discussion dealing with comparative modal incidents, it must be noted that valid incident comparisons between truck and rail modes are difficult to make. Each encompasses operations unique to the mode. For example, the motor carriers pick up, transport, and deliver hazardous materials packages; thus, they have a greater opportunity to note and report incidents. This is in contrast to rail operations that rarely include loading and unloading of boxcars and tankcars. This is done by shippers and consignees of the product who are not required to file incident reports. In addition, the number of individual packages transported in the highway mode is greater; therefore, the potential for an incident is greater.

From 1971 to 1979 there were 6254 injuries and 229 deaths from hazardous materials accidents. Of the injuries, 3941 were attributed to the highway mode and 2059 to the railroad mode (40). Of the deaths, 183 were highway-related and 42 were rail-related (40).

Property damage estimates for 1971 through 1979 for all modes other than pipelines are listed at about $109 million (40). The accuracy of the Materials Transportation Board (MTB) damage estimate statistics is in question, in part because the carrier reports must be filed within 15 days, when the full extent of property damage and injuries may not yet be known. The NTSB, which collects its data months later and counts all injuries, often reports damages that are greatly in excess of those reported by MTB. For example, in the Crestview, Florida, incident (see next section, Perspective) the MTB estimated damages at $0.136 million, whereas the NTSB estimate was $1.259 million (43).

A comparison of the last two years for which data are available--1979 and 1980--shows that in 1980 the total number of reported hazardous materials incidents, injuries, and damage estimates for all modes decreased from 1979 figures and the number of fatalities increased slightly as shown below (40):

Factor	1979	1980
Incidents	17,324	16,115
Injuries	941	619
Damages	$14,733 (million)	$10,707 (million)
Fatalities	18	19

A modal analysis of the 1980 data indicates that the largest number of incidents, injuries, deaths, and damage estimates is attributable to the highway mode. A rail-highway comparison is shown below (41):

	Rail		Highway	
Factor	No.	Percentage	No.	Percentage
Incidents	1327	8.2	14,484	89.9
Injuries	129	20.8	478	77.2
Fatalities	2		17	
Damages	$2,834 (million)	26.5	$7,324 (million)	48.4

Of the 19 fatalities that occurred in 1980, 6 accidents accounted for 14, or 73 percent of the deaths, and 5 of these accidents involved the highway mode (41).

The Inter-Industry Task Force on Rail Transportation of Hazardous Materials was formed in 1978 in response to concerns that arose from two major rail incidents that occurred within a two-month period in 1978 and resulted in 20 deaths. In mid-1979 the task force issued a report that indicated that, despite the fact that about 35 percent of all freight trains carry hazardous materials, only about 7.5 percent of all accidents involve a release of hazardous materials (13). In 1978, of the 1.1 million carloads of hazardous materials handled, only 151 carloads reportedly lost lading as a result of a derailment (13).

More than 90 percent of significant rail-related hazardous materials releases occur as a result of derailments (13). Rail and track geometry, truck components, and wheels are reported as the leading causes of derailment.

Modification of tank cars used in transporting hazardous materials--largely in the retrofitting of some 18,000 Class 112 and 114 tank cars (uninsulated, pressure tank cars constructed to DOT Specifications 112 and 114) with shelf couplers, head shields, and thermal protection to lessen the possibility of tanks being punctured by the coupler of the adjacent car in derailment--was required under Docket HM-144 rulemaking and has been under way for a number of years. Similar regulations are proposed for some 7000 Class 105 (insulated, steel-jacketed) tank cars. This action, together with technological improvements to decrease the likelihood of tank car breaching due to damage to top fittings or bottom outlets, is expected to offer more protection from damages caused by derailment.

Although rail incidents and accidents tend to receive the majority of the media coverage, truck transport of hazardous materials, as noted earlier, results in the greater number of incidents, accidents, and property damage.

Factors that contribute to accidents involving trucks that carry hazardous materials are those that contribute to trucking accidents generally: driver performance, vehicle and equipment performance, road geometry, etc.

Of the total number of accidents involving trucks that transport hazardous materials, a significant number were a result of train-truck collisions. The majority of the collisions occurred at passive crossings, which often were near hazardous materials storage or distribution terminals. The 310 accidents of this nature over the period 1975-1979 resulted in 35 fatalities, 206 injuries, and more than $8 million in property damages. A single truck-train accident investigated by NTSB in 1981 resulted in the derailment of 5 locomotive units and 24 cars, 1 fatality, and more than $2 million in property damages (21).

Perspective

The total number of fatalities resulting from hazardous materials transportation accidents is very small in comparison with other transportation-related fatalities. For example, there were only 19 fatalities resulting from hazardous materials accidents in 1980, whereas in the same year 51,900 lives were claimed by highway accidents; 1480 in marine accidents (including recreational boating); 528 in railroad accidents; 1375 in general aviation accidents; 33 in air carrier accidents; and 20 in pipeline accidents. Although these figures indicate the relative safety of hazardous materials transportation, the growth in the transportation of hazardous materials and the severity potential of a single hazardous materials accident must be carefully considered to frame the problem properly. It is estimated that the number of hazardous materials shipments will increase from 250,000/day to 500,000/day during the 1980s (24). New hazardous materials are being added daily to the thousands already in transport. The Occupational Safety and Health Administration (OSHA) estimates that 1 million compounds are manufactured yearly with 500 new compounds per year.

The increase in and added complexity of the traffic are only part of the reason for the concern about hazardous materials. If there were 500,000 shipments/day and the probability of an accident were one in a million, the accidents, especially if of some severity, would be too frequent for public acceptance. The severity of a single occurrence is an important and readily documented point, as the following incidents will exemplify.

- In 1970, the business district of Crescent City, Illinois, was almost completely destroyed by a boiling liquid expanding vapor explosion (BLEVE) of LPG after a railroad derailment.

- A derailment and series of explosions in Mississauga, Ontario, a suburb of Toronto, in November 1979 was coupled with a release of chlorine. Evacuated were 217,000 persons, including patients in three hospitals and six nursing homes. One-third of those evacuated in the 60-mile² area were unable to return home for a week (12). This incident, in which no lives were lost, is considered to be one of the worst chemical transportation disasters that has ever occurred. More than 1000 emergency personnel were involved in what was also one of the best-handled chemical disasters. Property damage estimates were placed at roughly $1.5 million. Overtime costs for emergency personnel were estimated at $0.75 million.

- In April 1979, 29 railcars, of which 26 carried placarded hazardous materials, derailed near Crestview, Florida. Chlorine gas and anhydrous ammonia formed a cloud that threatened a 300-mile² area downwind. This threat, lasting nine days, forced the evacuation of about 4500 people. Although there were no fatalities, 14 persons were injured. Damage was estimated at $1.3 million.

- A 1978 derailment near Youngstown, Florida, resulted in the release of the contents of a chlorine tank car. There were 8 deaths, 138 injuries, and more than $1 million in damages.

- In 1980 a locomotive collided with standing cars in a switchyard in Somerville, Massachusetts, a Boston suburb, and resulted in a punctured tank car that spilled 11,500 gal of liquid phosphorus trichloride. Water sprayed on the spill by local firefighters produced a reaction that caused clouds of acid vapor to form over the area. No fatalities occurred, but 23,000 people were evacuated and more than 400 people were hospitalized.

LEGISLATION AND REGULATION

Federal legislation affecting the movement of hazardous materials, toxic substances, and wastes has been promulgated since 1866. Beginning with the Act to Regulate the Transportation of Nitro-Glycerine, or Glynoin Oil, and Other Substances Therein Named (14 Stat. 81, July 3, 1866), some 64 such laws have been identified (23). The principal responsibility for regulating the transportation of hazardous materials rests with the federal government through the Commerce Clause of Section 8 of the U.S. Constitution. This has been broadly interpreted to include the movement of hazardous materials in both intrastate and interstate commerce.

The Transportation Safety Act (also referred to as the HMTA) authorized the Secretary of Transportation to regulate the safe shipment of hazardous materials. The basic regulations are found in 49 CFR 100-199. Additional statutory authority in Title 49 includes the Natural Gas Pipeline Safety Act of 1968, the Hazardous Liquid Pipeline Safety Act of 1979, the Dangerous Cargo Act of 1940, the Federal Aviation Act of 1979, the Ports and Waterways Safety Act of 1972, and the Federal Railroad Safety Act of 1970.

The U.S. Coast Guard (USCG) and EPA were authorized in the Clean Water Act of 1977 to act directly to eliminate a release, or a substantial threat of a release, of a hazardous substance into surface waters and to mitigate any resulting damage. Title 40 of the Code of Federal Regulations (CFR) includes the National Contingency Plan based on this Act, which includes broad powers to EPA to respond to a hazardous materials incident. The Resource Conservation and Recovery Act of 1976 also provided EPA with regulatory authority in hazardous waste and new chemicals, respectively. Related regulations based on these acts are also found in Title 40. The Comprehensive Environmental Response, Compensation, and Liability Act (CERCLA) requires notification to the Coast Guard's National Response Center in the event of a release of a reportable quantity (RQ) of a hazardous substance or a substantial threat of a release to the environment. Notification requirements of multimedia release, i.e., to air, water, and ground modes, are included for 612 substances. The purpose of this notification is to allow EPA or the Coast Guard to decide whether or not government response to the release is necessary (14).

Title 46 of the CFR codifies regulations for waterborne bulk cargos, and Title 33 codifies regulations for the safe handling of hazardous materials on commercial vessels and waterfront facilities. Title 23 regulates state programs for the cleanup of hazardous debris after accidents.

RESPONSIBILITIES

Federal Agencies

Department of Transportation

The HMTA authorized DOT to regulate the transportation of hazardous materials in commerce. The Secretary of Transportation may issue regulations applicable to both interstate and intrastate transportation of hazardous materials. The HMTA authorizes the Secretary to

1. Establish and revise criteria for handling hazardous materials;
2. Require hazardous materials carriers, shippers, and package-container manufacturers to submit biannual registration statements;
3. Grant exemptions to issued regulations;
4. Inspect records and properties relative to packaging, containing, and transporting of hazardous materials;
5. Provide both civil and criminal penalties for violations of the HMTA or its regulations;
6. Establish facilities and staff to evaluate risks and set up a central reporting and data system to facilitate hazardous materials emergency response; and
7. Conduct a continuing review of all aspects of the transportation of hazardous materials to recommend steps to ensure safety and to prepare an annual report to Congress.

DOT is also empowered, under various other pieces of legislation, to regulate the transport of hazardous materials by individual mode. For example, DOT regulation of rail transport of hazardous materials is provided by the Federal Railway Safety Act [45 USC 331 (1976)] and highway transport by the Department of Transportation Act [49 USC 1655(e)(6)(c) (1976)] and under the Motor Carrier Act of 1980 (94 Stat. 793), which requires motor carriers to carry liability coverage and to compensate injured parties for personal injuries (medical costs and impairment of earning capability) and replacement and/or restoration of property or natural resources destroyed in hazardous materials accidents. The minimum levels of financial responsibility requirements for carriers were superseded by Section 406 of the Surface Transportation Assistance Act of 1982 (96 Stat. 2097), which (a) expanded the authority of the Secretary of Transportation to require minimum levels of financial responsibility for motor vehicles transporting hazardous materials, hazardous substances, or hazardous wastes by foreign carriers in the United States engaged in foreign commerce; (b) required motor carriers domiciled in any contiguous foreign country to carry on board each vehicle it operates in the United States evidence of financial responsibility; and (c) expanded the applicability of the minimum levels of financial responsibility requirements to include motor vehicles that have a gross vehicle weight rating (GVWR) of less than 10,000 lb when transporting certain hazardous materials.

A major objective of the HMTA was to consolidate previously fragmented regulatory and enforcement powers involved in the transportation of hazardous materials. In implementing the Act, the Secretary of Transportation delegated authority to the MTB within RSPA.

MTB is primarily responsible for coordinating the efforts of the four modal administrations to provide comprehensive oversight of the hazardous materials industry. Its responsibilities include the development and issuance of regulations and exemptions that govern the transport of hazardous materials by all modes (with the exception of bulk transport by water). These regulations cover shipping and carrier operations, packaging and container specifications, hazardous materials definitions, and communication of hazards.

MTB is also responsible for enforcement of regulations other than those applicable to a single mode of transport. Enforcement of regulations applies to manufacturers, reconditioners, and retesters of DOT specification containers and to multimodal shippers of hazardous materials. The other modal administrations--USCG, Federal Aviation Administration (FAA), Federal Highway Administration (FHWA) through its Bureau of Motor Carrier Safety (BMCS), and Federal Railroad Administration (FRA)--originate, issue, and enforce dedicated hazardous materials regulations for each of their respective modes of transportation. MTB is responsible for pipeline commodity transport.

MTB sponsors a number of training and education programs designed to foster compliance with hazardous materials transportation regulations. Such programs cover training for rail, motor vehicle, and maritime operations relevant to the movement of hazardous materials and develops hazardous materials training materials for both public and private-sector parties. DOT's Transportation Safety Institute in Oklahoma City routinely conducts in-depth training courses for government and industry personnel in all phases of hazardous materials transportation enforcement.

Coast Guard

The USCG issues and enforces regulations related to the maritime transport of bulk shipments. It is responsible for federal activities required to mitigate oil and hazardous materials threats to the shoreline of the Great Lakes, navigable rivers, and coastal areas, and operates the National Response Center, located in Washington, D.C.

Established in 1974, the Center receives initial reports of spills of hazardous materials and substances and oil, as required under the provisions of the Clean Water Act, and provides facilities, communication, information storage, and other needs for coordinating emergency response. The Center's data base, Chemical Hazards Response Information System (CHRIS), contains information on the characteristics of some 900 chemicals and can be used to develop predictions of the behavior, movement, and human and environmental effects of chemicals. The Center operates 24 h/day and works cooperatively with CHEMTREC, a similarly structured private-sector emergency center (see section on Industry, p.12).

The USCG and EPA jointly share lead responsibility for on-site response. This responsibility is exercised locally by on-scene coordinators (OSC) supplied by both agencies. These coordinators are empowered to countermand activity at the scene and to redirect the response, if necessary--a rare occurrence.

Environmental Protection Agency

The Resource Conservation and Recovery Act of 1976

requires EPA to set standards for transporters of hazardous wastes. These standards cover recordkeeping, labeling, and compliance with a manifest system for origination-to-disposal management of hazardous wastes. EPA regulations must be coordinated with DOT and be consistent with DOT regulations concerning any materials subject to the HMTA (23).

The Toxic Substances Control Act of 1976 authorized EPA to regulate the manufacture, processing, and distribution in commerce of chemical mixtures and substances, the use or disposal of which may present an unreasonable risk of injury to health or to the environment. It does not authorize EPA to regulate the transportation of such substances, but it does intend that there be coordination with agencies such as DOT that could be used in achieving the Act's goals (23).

The Clean Water Act of 1977 gives EPA the authority to respond to a hazardous substance incident if (a) the pollutant is identified as a hazardous substance, (b) there is a discharge or substantial threat of a discharge of the substance into U.S. waters, and (c) the discharge or potential discharge involves such a quantity as to create an unacceptable threat to the environment or to the general public.

The satisfaction of these conditions is confirmed by an EPA OSC. Many incidents that meet the above criteria will occur during hazardous materials transportation. Once a discharge, or a potential discharge, to the environment exists and is determined by the OSC to be an unacceptable threat, EPA is authorized to respond. CERCLA extends response authorization to multimedia environmental releases of RQs.

EPA has divided the United States into 10 regions. Each region has an office with an OSC available. There are Regional Response Teams and an Environmental Response Team to mitigate spills.

Department of Energy

The U.S. Department of Energy (DOE) is the coordinating agency for the Interagency Radiological Assistance Plan (IRAP), which provides rapid and effective assistance in handling radiological incidents. DOT is a participating agency along with the U.S. Nuclear Regulatory Commission (NRC) and about 10 other agencies. To implement IRAP, DOE has divided the nation into eight regions, each with a Regional Coordinating Office. DOE also operates the Joint Nuclear Coordinating Center in Albuquerque, New Mexico, which has special expertise with regard to nuclear weapons.

Federal Emergency Management Agency

The Federal Emergency Management Agency (FEMA) is an independent federal agency. It was established in 1978 and has overall responsibility for coordinating the federal response to disasters. However, its role in the transportation of hazardous materials has not been fully clarified. FEMA operates the National Fire Academy, which provides training in fire service technology and in disaster planning for hazardous materials incidents. The training program operated by the National Fire Academy is thought by many to be the model to be emulated in developing appropriate emergency response training programs for local-level fire and emergency personnel.

Nuclear Regulatory Commission

The NRC is the designated agency to enforce nuclear materials regulations. The Nuclear Waste Transportation Safety Act of 1979 amended the HMTA to

1. Define DOT and NRC responsibilities for nuclear materials shipment regulation, with DOT primarily responsible;
2. Require DOT to develop a national plan for responding to radioactive materials transportation emergencies;
4. Provide grants to states to study safety and logistics for nuclear material shipments to disposal facilities; and
4. Require DOT to study and establish shipping standards for foreign nuclear material received by the United States.

The NRC has divided the United States into five regions. An office for each region is available for assistance in an emergency.

National Transportation Safety Board

The NTSB investigates "selected" hazardous materials accidents and issues reports on them. It also make recommendations to federal transportation agencies with regard to technological and institutional measures that would enhance safety in all modes of hazardous materials transportation.

State and Local Governments

State and local government responsibility for the safe movement of hazardous materials is fragmented and varies widely from locality to locality. A 1978 state survey made by the American Association of State Highway and Transportation Officials found that hazardous materials transportation regulatory authority was vested in 56 agencies within 41 states. Authority was vested in the transportation department in 21 states, in public service agencies in 15 states, in safety agencies in 10 states, and in environmental agencies in 10 states. A DOT study conducted about the same time reported that 97 agencies in 51 jurisdictions held such authority or responsibility (4).

The federal hazardous materials regulations had been adopted completely by only 20 states by early 1981. A few states had no rules at all and the remainder had either partially adopted the federal regulations or had similar rules of their own (41). Through state contractual agreements for technical and financial assistance, DOT encourages the adoption and enforcement of the federal standards. Uniformity of regulations is seen by most as the key to the development of a sound and consistent hazardous materials transportation safety program.

In general, the following types of agencies within a state have, to some extent, responsibility and/or authority for hazardous materials transport: Department of Transportation, Public Service Commission, Motor Vehicle Department, State Police or Highway Patrol, Environmental Agency, Civil Defense, Health Department, and State Fire Marshal. At the local government level, involved agencies include police, fire, health, civil defense, and traffic engineering agencies.

Fragmentation of responsibility and the lack of a clearly designated lead agency have resulted in a lack of communication between state and local government agencies and a failure to define roles and develop enforcement, compliance, training, and incident-response-capability policies and programs. To this failure to define roles and develop uniform regulations and programs must be added increased public sensitivity to hazardous materials transport. As a result of these factors, the number of restrictions imposed by local governments to prohibit or to regulate the movement of hazardous materials through their jurisdictions is likewise increasing. Such restrictions include requiring prenotification of shipments, time of day and routing constraints, and bans on specific commodities. These interjurisdictional problems are detailed in Part 4 of this report.

Industry

Responsible industry interest in the safe transport of hazardous materials and in the mitigation of emergency situations is most readily exemplified by the educational, training, and emergency-response programs of individual shippers and carriers, as well as by services provided by groups such as the Association of American Railroads' Bureau of Explosives (BOE), the American Trucking Associations, Inc. (ATA), the Chemical Manufacturers Association's CHEMTREC, the various specialized chemical groups such as the Chlorine Institute's CHLOREP, and membership organizations such as the Hazardous Materials Advisory Council (HMAC). Each of these is briefly described below.

Association of American Railroads, Bureau of Explosives

The BOE is headquartered in Washington, D.C., and has 19 field inspectors (16 in the United States and 3 in Canada). BOE also operates a chemical laboratory in New Jersey. BOE "tests and classifies for transportation purposes those materials considered hazardous under government regulation and makes recommendations covering packaging, labeling, and handling en route" (3). BOE works with member industrial groups as well as federal and military agencies, conducts safety audits of rail facilities and production/shipping sites, and assists railroad personnel in the handling of derailments where hazardous materials are involved. Their manual, *Emergency Handling of Hazardous Materials in Surface Transportation*, is widely circulated to shippers and local authorities (6).

American Trucking Associations, Inc.

The ATA Safety and Security Department has developed several programs designed to assist the trucking industry in meeting its obligation to safely transport hazardous materials. Coordinated under the direction of its Hazardous Materials Committee, ATA has developed a comprehensive motor carrier training program, including a supervisor's reference manual, a compliance seminar, and an employee audiovisual training package.

ATA has also developed an emergency response training program designed to enhance the knowledge of both police and fire officials in regard to hazardous materials and trucking operations. These emergency-response training programs are being conducted throughout the country by ATA's affiliated state trucking associations.

Chemical Transportation Emergency Center

CHEMTREC was established in 1971 by the Chemical Manufacturers Association. It is a toll-free telephone information hot-line that provides immediate and sustained communication with individuals at the site of a hazardous materials transportation emergency. The CHEMTREC data base contains information on the hazardous properties of some 18,000 chemicals or chemical classes. It also has extensive information on some 900 chemical companies, including their emergency response teams, and on ownership or leasing of tank cars; thus, it is likely to know whose product may be in which car.

On receipt of a call, usually from a carrier or emergency services personnel, the CHEMTREC communicator ascertains the location of the incident; the product(s) involved; identity of manufacturer/shipper, carrier, consignee; number of truck or railcar; and conditions prevailing at the scene. The communicator provides emergency information and information about the specific hazards involved. CHEMTREC then notifies the shipper, who, in turn, assumes responsibility for the response. By using a "teleconferencing bridge" developed by the Bell System, CHEMTREC can simultaneously link all affected parties, including the National Response Center.

Since 1980, CHEMTREC has been officially recognized by DOT as fulfilling the requirement of the HMTA that such an emergency response center be created. Under the provisions of a Memorandum of Understanding, dated March 13, 1980, CHEMTREC is required to notify the USCG National Response Center of "significant" hazardous materials transportation incidents. "Significant" is defined as those incidents that have, or have potential for, causing considerable harm to the public or the environment. Of the total number of incidents reported in a given years, only about 1 percent meets this reporting requirement.

Beginning in 1978, CHEMTREC, together with the Chemical Manufacturers Association, sponsored workshops for chemical industry employees to improve their emergency response capabilities. These workshops are organized and presented under the guidance of the CHEMTREC Advisors' Group. Instruction since 1980 has been provided by the Texas Engineering Extension Service of Texas A&M University, with support from chemical industry, carrier, and other personnel.

Chlorine Emergency Plan

CHLOREP is operated by the Chlorine Institute and is composed of 28 chlorine-producing companies that operate 62 regional emergency response teams in the United States and Canada. Within CHLOREP the concept is the same as it is for similar industry groups. The member company closest to the scene of the accident will send a task force, regardless of whose product is involved. Experience indicates tht the increased speed of response to an accident more than makes up for the potential problem of personnel from one company dealing with the product of another company (31).

Hazardous Materials Advisory Council

Cross-cutting all the issues and actors involved in the transportation of hazardous materials and wastes, HMAC is a non-profit, Washington, D.C.-based membership organization whose members represent carriers (all modes), shippers, container manufacturers, and trade associations. Through its seminars, conferences, and training courses, HMAC serves as a forum for the exchange of information on hazardous materials transportation legislation and regulations at the national and international levels.

PART 4

Discussion of Committee Findings and Recommendations

Committee Findings and Recommendations

FINDING 1: NEED FOR LEAD FEDERAL AGENCY

There should be a single lead federal agency responsible and accountable for the safe transportation of hazardous materials.

Hazardous materials transportation activities and responsibilities are fragmented among various organizational entities in DOT, EPA, and FEMA. In DOT, the MTB is the designated coordinator; however, the consensus of the experts assembled at Williamsburg is that the designation is not practiced so that the transportation community recognizes the existence of a single lead agency. It is difficult to perceive a single focus and direction in hazardous materials transportation and regulation because (a) modal regulations are managed in different organizational locations; (b) enforcement is separated from regulation management, as it is between MTB and BMCS; (c) multiple agencies are involved in emergency response; and (d) state or industry officials have no clearly designated point of contact in the federal framework. It is assumed that DOT is the lead agency. The finding of the committee that there must be such an agency implies that, regardless of existing statutes, the agency is not fulfilling the role.

In order to safeguard the public interest and to ensure the safe and efficient movement of hazardous materials vital to the national economy, there must be a single, strong, responsible, and accountable federal lead agency to provide the leadership required for coordination of the effort for uniformity in regulation, enforcement, research and development, emergency response, and training. Such leadership must emanate from a vantage point that perceives the total picture. It must also understand the local tendency to respond to an undesirable hazardous materials incident by regulating to prohibit any chance of what is now viewed as an eminent rather than a distant threat.

The federal agency must first define the problems that these other people are responding to or enforcing against. The federal agency must define the problems, determine their causes, assign a certain priority to them, and resolve them through the issuance of nationally uniform regulations or other corrective measures. Once the federal effort is properly aimed to resolve the problem, other bodies can carry out the functions of meeting, enforcing, and responding. The federal agency, however, must be the starting point for all other parties' actions.

The mere designation of a lead federal agency is insufficient. There must be oversight mechanisms designed to assure that the agency performs its designated functions, to measure that performance, and to assure that organizational and management practices do not hinder the efforts of dedicated officials. The agency must also be designed and organized for maximum visibility and accessibility and for the ability to identify and respond to the current, continuing, and episodic problems in hazardous materials transportation.

DOT, rather then FEMA or EPA, is the appropriate agency to be given this responsibility. FEMA is a young agency with primary responsibility for civil defense against disasters. Rarely does the transportation of hazardous materials result in a catastrophe because the bulk of the effort in this field focuses on prevention. Similarly, EPA is concerned with discharges to the environment. Most hazardous materials are transported without releases to the environment. The greater part of the effort to protect the public is intensely committed to transportation technology and management. This is as it should be. DOT should have both the authority and the responsibility for safe transport of these materials. It cannot contain all federal actors, but it can be given authority to use those actors in the most effective way possible. This should include using them for the most effective mitigation of the consequences of an undesirable event occurring during hazardous materials transportation.

The committee concluded that DOT can and should perform the role of a single lead agency and can be designed to interact effectively with state and local governments and with industry.

RECOMMENDATION 1: LEAD AGENCY

There should be a lead federal agency for the development of hazardous materials transportation policy and procedures. This agency should be the Department of Transportation.

The conference committee recognizes that this is a departure from current laws and practice in that Congress has given other agencies authority in various aspects of hazardous materials transportation. Nevertheless, DOT should be designated to resolve disputes, clarify boundary relationships as previously mentioned, and lead the regulatory efforts for hazardous materials transportation.

DOT should reexamine its organizational structure to provide recognizable points of authority and contact for increased interagency, intra-agency, and agency-industry coordination. The organizational entities should be structured to meet the current, continuing, and episodic problems in hazardous materials transportation. These needs include, but are not limited to, training, regulation, enforcement, emergency response, research and development, and agency-industry coordination.

DOT should take a strongly proactive stance in discharging its lead-role responsibilities. Specifically, it should establish, convene, and confer regularly and often with a body of officials representing regulated industry and federal, state, and local governments as it develops all hazardous materials transportation policies and procedures.

The specific issues on which the body established by DOT should address and advise all responsible parties include, but are not limited to, the following:

1. A definition of purpose of the federal regulatory program that, while acknowledging limits on resources, establishes measures of safety effectiveness;
2. Consistent and uniform training, where feasible, of all appropriate parties (private and government) throughout the country;
3. Simplification and clarification of all regulations, with emphasis on performance standards and guidelines where design specifications and mandated procedures now exist;
4. Preparation and publication of a long-range regulatory plan;
5. Role definition (emphasizing the value of uniformity of procedure and action) for local, state, and federal officials involved in regulation, enforcement of regulations, and responses to incidents;
6. Enforcement practices that recognize the critical importance of industry's civil liability and the dominant role of state and local officials as instruments of federal policy; and
7. Emphasis, where emergency response is concerned, on the three priority issues: (a) planning, which must be coordinated by a single agency at each level of government; (b) training, which should be designed and conducted in accordance with the needs of responders; and (c) authority, which involves the designation of an official responsible for incident management in each jurisdiction.

The hazardous materials regulations should be reduced in complexity and volume. Among other things, this can be accomplished through a reduction in the number of regulated materials, drafting of provisions to enhance training and user comprehension, publication of introductory instructions on how to use the regulations, development of charts and graphs where feasible, reliance on accepted-practice guides instead of detailed specific requirements, minimizing of cross references and exceptions, and avoidance of legalistic phrasing.

As part of establishing a strong lead role, DOT should improve its rulemaking process to stimulate greater public involvement by preparing and publishing a long-range regulatory plan, using more advance notices, providing adequate comment periods, writing more explanatory preambles in lay terms, engaging in deeper investigation of technical and operational facts, soliciting input from affected parties, correlating each project with the overall plan, standardizing reasonable effective dates, and starting and concluding projects expeditiously.

Risk assessment and historical data should be used to identify regulatory and enforcement priorities, as well as areas where these efforts may be reduced. There should be functional implementation of regulation for those substances that are in a quantity and form that, when transported, may pose an unreasonable risk to health and safety or property.

DOT should establish an enforcement policy that (a) focuses on regulations that have the greatest correlation with assuring the safety of the greatest number of people, (b) recognizes limitations on government and private resources and encourages allocation of those resources to achieve the greatest safety benefit, (c) provides mode-to-mode and case-to-case unformity and consistency in procedures and penalties, and (d) relies on non-federal governments for implementation. The agency should publicly identify those specific regulatory provisions that benefit most from the involvement of industry, federal inspectors, and state and local officials engaged in implementation of nationally uniform rules. To the extent possible, the duties of enforcement and training should be performed at the non-federal level.

DOT should begin to identify and estimate the probable causes and contributing factors in hazardous materials transportation accidents. This will require an interdisciplinary approach to improve the understanding of accident dynamics, data acquisition, and analysis.

FINDING 2: SAFETY GOAL

There is no agreed-on hazardous materials safety goal within the government-industry complex. In large measure this is due to a lack of defined purpose underlying the federal regulatory program. Lacking such purpose, many regulations as they exist today are based on special-interest concerns, crisis reactions, and perceived needs rather than a more rational risk management approach. Further, this piecemeal evolution of hazardous materials transportation regulations has resulted in a pervasive and unnecessary complexity that may in fact discourage the very safety it proposes to assure.

Except for the following vague legislative policy statement preceding the HMTA, there is no written statement of purpose for the existing hazardous materials transportation regulatory program:

> It is declared to be the policy of Congress in this chapter to improve the regulatory and enforcement authority of the Secretary of Transportation to protect the Nation adequately against the risks to life and property which are inherent in the transportation of hazardous materials in commerce (49 USC 1801).

The experts meeting at the Williamsburg conference demonstrated a surprising diversity of views on the apparent purposes of the federal program. However, it is not at all certain from any existing documentation who or what the program should be protecting, from what types of harm, through what mechanisms, and with what considerations.

The federal rules long predate the current HMTA. They have evolved over most of a century, as did the U.S. transportation and chemical industries. The effort has never received substantial administrative attention, to the end that most actions over this long period have been designed only to keep the existing program running and to make modest adjustments to small segments of it to account for individual problems arising at the moment.

It appears that the origin of the existing program lies in the commercial and legal relationship between hazardous materials shippers and common carriers who, under the common law, were obligated to accept the freight and were the insurer not only of that cargo but also of other cargo entrusted to them. Common carriers' obligations were legally based on awareness of the character of the cargo, and thus there was a basic shipper obligation to inform the carrier of the hazards inherent in the shipment.

The communication of hazard information to the common carrier was done with shipping documents and various markings and hazard warnings on packaging. These communication requirements were augmented by a shipper obligation to package hazardous materials so that they might be handled properly and safely in transportation. The safety of carrier personnel and the protection of adjacent cargo insured by the carrier were the objects of these measures. These original shipper obligations, initially applicable to common carriers by rail, were codified by the U.S. railroad industry. Early in the 20th Century, the carrier function was adopted as a criminally enforceable government regulatory program. For many decades, however, until the 1960s, the railroad industry played a more substantial role in the development of these regulations than the government, which adopted and enforced them.

With the evolution of highway, air, and water transportation modes in the United States, the initial regulatory function addressed to rail carriage was expanded to accommodate them. So, too, as different hazards were identified, or new materials were developed, these "dangerous articles" were incorporated into the existing system. Dramatic improvements in packaging, with the development of steel drums, cylinders, plastics, and the like, led to appropriate adjustments to packaging requirements imposed by the regulations on hazardous materials shippers.

For these apparent original purposes of the program, it was adequate and, for these purposes, perhaps remains adequate. It is the conclusion of the conference committee, however, that the purposes of the program may have changed, or may need to change, because of changes in society over the last 100 years. Reasons for change include an increase in population densities, more complex transportation systems, enhanced environmental concern, greater knowledge of the health effects of certain materials, substantially increased public interest and awareness evidenced through media coverage and public-interest group and local government activities, and greater technical competence of all parties.

These increases, however, are countered by a decrease in available resources to cope with the problems. It is unacceptable to simply add to the program every time a new concern is voiced. A systematic, efficient approach to the area is absent, but clearly is necessary.

RECOMMENDATION 2: PURPOSE STATEMENT

Virtually all of the recommendations of the conference committee are based in part on this second finding. One of the key recommendations, however, is that DOT should take a lead role in conferring with representatives of regulated industry and all levels of government to define explicitly and in writing the safety goals of the federal regulatory program.

The representative group need not have the statutory mandate of the Congress or have the level of formality envisioned by the Federal Advisory Committee Act (5 USC App., Sec. 1), but it is clear that the formation of a communications link among DOT, other federal agencies, industry, and state and local governments is essential. Such a group would have multiple functions, as addressed in the basic findings and recommendations of the conference committee. It would serve as an identifiable conduit for information and views from all parties to DOT, and from DOT back to such parties, and with greater directness than Federal Register publications.

The committee found a gap in communication between all parties, running both to and from DOT, and this advisory body could do more to bridge that gap than any existing mechanism. It is envisioned that at some time there may be no continuing need for this communications link, but at this time there is a vital need for it because of the inability of DOT to communicate with all affected parties and the perception of those parties that incoming communications are not heard.

One of the initial functions of this DOT-sanctioned group would be the identification of the goals of the current program. The goals of the program should be encapsulated in a purpose statement. Until all interested parties know why the federal program exists, there is no means by which to measure its effectivenss or to determine the resources necessary to achieve the stated ends. Officials in related federal agencies and at other levels of government are unable to perceive their own responsibilities without a clear understanding of the scope of the DOT program. Administrators of the DOT program itself are unable to determine their priorities or develop long-range plans without an agreed-on statement of the goals of their efforts.

The purpose statement should be viewed as a charge, or mission, for the program. It also should be viewed as a limitation on that program, so that any activities not within the scope of the purpose statement are deemed inappropriate and are abandoned or not undertaken.

The more general the purpose statement, the less helpful it will be. It must be sufficiently general, however, to accommodate circumstances in the future that are unknown at this time. The Williamsburg conference revealed substantial diversity of views on this issue. It is not the intent of the conference committee to draft the purpose statement. As noted above, that should be done in a public forum with participation by all affected parties. For the sake of illustrating this recommendation, however, the following purpose statement has been consolidated from several developed at Williamsburg:

> It is the objective of the national regulatory program to assure the safe and effective transportation of hazardous materials, through development of feasible requirements and compliance incentives that promote domestic and international harmony and transportation efficiency, while minimizing the risk to life, health, and the environment. Since the resources of society are limited, regulation should be designed to reduce the significant potential for simultaneous harm to many persons or highly valuable natural resources. The Secretary of Transportation is the lead national official charged with this responsibility, but the Secretary shall give full consideration to the views of affected parties, and shall provide technical guidance to state and local governments in the implementation and enforcement of nationally uniform hazardous materials transportation regulations.

The key terms in this example of a statement include the following:

1. "national"--This program is the one for the entire country and is not limited to DOT or to interstate commerce.
2. "safe and effective"--Safety is the key goal, and in the event of conflicts between what is appropriate for the environment and what is appropriate for human safety, safety takes precedence. Combining this term with effectiveness means that each regulatory effort should be assessed to assure that it achieves the ends intended with the least expense.
3. "feasible requirements"--Here, again, the economic and practical reality of regulation should be considered.

4. "compliance incentives"--The regulatory program should not only be viewed as a negative series of restrictions, but it should also encourage voluntary actions to achieve safety.

5. "promote domestic and international harmony and transportation efficiency"--Disruptions to commerce that result from inconsistent regulatory programs developed by other bodies, whether private or government, is an appropriate concern for DOT, which justifies administrative initiatives to achieve harmony here and abroad.

6. "while"--The development of feasible requirements and promotion of harmony must be done concurrently with reduction of risk.

7. "minimizing the risk"--This should not be a program of achievement of safety at all costs. Risks should be reduced, but it should be recognized that fuels and other vital materials cannot move in a totally risk-free manner.

8. "life, health, and the environment"--Hazards in transportation to health and the environment are subject to regulation as well as articles that bear the potential for immediate human injury.

9. "resources of society are limited"--Getting the best program for the least expense is an affirmative obligation of the agency.

10. "reduce the significant potential"--Reduce, not eliminate, is the key. Further, the potential should not be imaginary or inconsequential, but should be significant.

11. "simultaneous harm to many persons or highly valuable resources"--The program should focus on the biggest threats to the most people, or that property or environmental resource that has the greatest value. Specifically, attention should be given to standards for tank cars, cargo tanks, and bulk containers and vessels in all modes of transportation. Devotion of limited administrative and public energies to the regulation of low-risk matters is inappropriate.

12. "Secretary of Transportation"--The hazardous materials transportation regulatory function should remain in DOT.

13. "lead national official"--DOT should take the affirmative leadership role in this field, rather than having the transportation program fragmented horizontally at the federal level or vertically throughout all levels of government.

14. "responsibility"--This program is a duty, not a discretionary option.

15. "full consideration"--The regulatory program should affirmatively seek the views of all affected parties, not just those who take the initiative to follow the Federal Register and comment.

16. "technical guidance to state and local governments"--This, too, is an affirmative duty, not just an optional activity.

17. "implementation and enforcement"--The guidance should cover all procedural, practical, and administrative steps necessary to enforce standards at multiple levels of government.

18. "nationally uniform hazardous materials regulations"--No guidance or assistance should be proffered for the implementation or enforcement of any requirements that disrupt the safety achieved through nationally uniform requirements.

Such a purpose statement, much like a constitution or similar document, is a basic structure that will take greater form over time through experience and interpretation. It is anticipated that periodic amendments to it may be necessary.

Whereas many of the elements of the purpose statement may appear obvious and one might assume that they are part of the existing program, the conference committee has found that they are not a part of the existing program, hence the need for this recommendation.

FINDING 3: ROLES OF GOVERNMENT

Coupled with the second finding--that there is no agreed-on hazardous materials transportation safety goal within the government-industry complex--is the third finding that the relative roles of federal, state, and local governments never have been clearly and understandably defined. This has led to interjurisdictional confusion and conflict. In addition, no developed concept exists on how best to achieve compliance with those measures that will provide the greatest safety.

Relative Roles of Government

Regulation of local safety is traditionally a matter of local police powers. In the highway mode, where construction, maintenance, and use of roadways often are the responsibility of the local community, the operation of vehicles on those highways has been considered a matter of local concern. Preemption by the federal government is not to be assumed. In the case of Interstate highways, the reasons to follow this rationale are less persuasive.

In the field of hazardous materials transportation, Congress has expressed a concern for the traditional divisions of responsibility between the federal government and the states.

The HMTA, among other things, was developed to cure a fragmentation of regulatory activities in the field, carried out by a variety of agencies, levels of government, and private parties. DOT is made up of several independent operating administrations that address issues in their respective modes--air, water, highway, and rail transportation. The original DOT statute, which gathered hazardous materials regulatory authority from several diverse sources and put it into a single agency, obligated the agency to keep these authorities separate. Thus, any action affecting all the modes, such as a change in packaging, required the signature of four different officials, as well as the staff input and review of each mode prior to that conclusion.

In the legislative demand for greater efficiency was a demand for uniformity, and this also led to enactment of a qualified preemption provision. Any state or local enactment that is deemed "inconsistent" with the HMTA or the regulations issued thereunder is preempted as a matter of law under 49 USC 1811(a). In explaining this provision, the U.S. Senate Commerce Committee reported that it "endorses the principle of Federal preemption in order to preclude a multiplicity of State and local regulations and the potential for varying as well as conflicting regulations in the area of hazardous materials transportation" (38).

Broad public concern with the transportation of hazardous materials, including hazardous wastes, has led many state and local governments to enact their own regulations on this transportation.

The conference committee finds a strong safety value in having nationally uniform requirements, in addition to any commercial benefits. As noted throughout this report, one of the key obstacles to voluntary compliance is the complexity of DOT's regulations. Although implementation of measures such as those recommended in this report could reduce that complexity, it is understood that the subject matter itself is complex and that only so much clarity may be achieved.

In a complex field involving carriers, drivers, and freight crossing many jurisdictional lines throughout the country, where peoples' levels of understanding and capability span a broad spectrum, any additional confusion is detrimental to compliance. Having transportation rules that are unique to a particular community or several different communities is destructive to compliance with the federal system. There may be certain areas of action, however, where the benefits from the local rule outweigh the impairment to uniformity.

There are several areas that DOT has proclaimed to be exclusively federal and where no state or local action is authorized. One of these is hazard communications, which under the federal system take the form of shipping papers, package markings, package labels, and vehicle markings and placards. Another is in hazardous materials containment, including all aspects of packaging from the basic container to all valves and fittings on it. A third is the field of hazard terminology and definition, where DOT has declared as preempted those local definitions of hazardous materials in transportation that differ from the federal terms. The last area is in written accident reporting, which is done under a uniform national reporting system, the results of which are available to any community.

The HMTA does not preempt all state and local regulatory actions. Valid areas of local activity deal primarily with the highway mode of transportation, in emergency response and in the operation of vehicles in that mode. Examples of the kinds of regulations frequently adopted at the local level include vehicle registration, permitting, routing, curfews, prenotification, and bans. Unless such local rules conflict directly with a given federal rule, the measure is whether the local rule poses an obstacle to accomplishment and execution of the purposes and objectives of Congress in establishing the federal program.

One of the principles inherent in the federal program is avoidance of unnecessary delay. DOT and the courts have found local requirements that unnecessarily delay transportation to be preempted. An example is a requirement that a written permit be obtained for each trip by each vehicle, which could not be applied for until the vehicle was loaded. A key factor in the impact of many local operational rules is the shifting of regulated traffic from the enacting jurisdiction into other jurisdictions. Here the concern is with one community attempting to minimize its exposure by increasing the exposure of its neighbors. The important elements of any evaluation of such traffic shifting are that (a) the routes that will be used in fact are safer than those that would have been used, and (b) implementation of any traffic-shifting requirements be coordinated with all affected jurisdictions.

DOT's views in these matters are not self-enforcing, nor has the agency taken the initiative to seek judicial enforcement of its conclusions. For several reasons it is unlikely any court will take DOT's inconsistency rulings at face value without further examination. The first is that DOT's inconsistency ruling procedure is too informal to constitute a very good fact-finding mechanism. Courts determine facts through the examination of evidence under careful rules of admissibility and testimony under oath that is subject to cross-examination. In contrast, DOT gathers background information through a Federal Register request for comments. Any person may say anything, and there is no mechanism by which one may assure the completeness of the information or its veracity. Accordingly, courts that are asked to implement a DOT ruling are more likely to use that ruling in their own examination of the issues as technical guidance from an administrative body entrusted with the implementation of the statute rather than as an order that warrants full judicial deference.

If any state or local jurisdiction admits its rule is inconsistent, or it is found to be so by DOT or by a court, then the state may apply to DOT for a "non-preemption" determination under 49 USC 1811(b). In explaining this process, the Senate Commerce Committee referred to a need to accommodate "certain exceptional circumstances" that may necessitate "immediate action to secure more stringent regulations" (38). The courts may not waive preemption; this is a power unique to DOT. The local rule must be more stringent and may not unreasonably burden commerce. However, not every inconsistent local rule that is more stringent and does not unreasonably burden commerce will be authorized by DOT because DOT must weigh the damage of the inconsistency against the benefit of greater stringency in deciding whether the local rule may stand. This procedure has remained virtually unused but provides a means of approval of rules that might otherwise be invalid.

As may be evident from the preceding discussion, most of the definition of the relative roles of federal, state, and local governments has come through conflict. Although from the resolution of disputes comes a better understanding of relative roles, this is an extremely inefficient, unpredictable, and expensive way to proceed.

There has been virtually nothing done in a positive fashion to define and coordinate the various roles of different levels of government in the development, promulgation, interpretation, and enforcement of hazardous materials transportation regulations.

Concepts of Compliance to Achieve Safety

The following discussion addresses nine general issues that the Steering Committee feels must be considered in order to develop concepts of how best to achieve compliance with measures that will provide the greatest safety.

Regulations and Enforcement

Enforcement is an essential element of an effective regulatory process. Close coordination and cooperation between those who select regulatory actions and those who enforce them are vital. If the intent of a regulation is to solve a safety problem, then enforcement of that regulation is essential. Requirements highlighted through enforcement are stressed in industry compliance efforts, and this energy should be harnessed to achieve the intended safety objective. Enforcement should be coordinated with regulation and, after a full period to allow for implementation and after enhancement of awareness through education, vigorous enforcement and publication of enforcement efforts should be undertaken.

It is clear that DOT's current enforcement program needs improvement. Centralization of the function, or at least unification of procedures, appears warranted so that the penalty suffered by a respondent does not vary due to the affiliation of the inspector.

In enforcement, as in other aspects of regulation, some guidelines are necessary to preclude arbitrariness and inconsistency. Today, penalties may vary by mode of transit, by the procedural avenues selected, and by the attitudes and perceptions of personnel assigned to the case. The current system appears too subjective and without consistent relation between the penalty and the gravity of the offense.

An enforcement program that thrives on cases that are easy to prove, regardless of their correlation with safety, is a disservice to the public. As a revenue-producing measure, it is inefficient and cannot be justified as a safety program. Focus on minor requirements and assessment of small fees in the hope that the respondent will not undertake the expense of resistance also disserve the public.

Penalties and fines should correlate with the severity of noncompliance. Penalties should fall most heavily on those who have the most severe and frequent violations. This also assumes that the enforcement program and the penalties relate to the levels of hazard of the material and of the risk--i.e., for high levels of risk there should be higher levels of enforcement.

A vigorous enforcement effort that seeks significant penalties to deter future noncompliance with significant requirements necessarily results in requirements that are more soundly based and more easily understood. The current program, which avoids the hard cases because the regulations are unclear, is not serving one of its vital functions--the achievement of greater public safety through improvement of the regulations.

The elements of the purpose statement for the national hazardous materials regulatory program should be reflected in the enforcement and compliance system. As a regulatory technique, the issuance of detailed regulations implies the necessity of an enforcement program. Without such a program, a system of voluntary compliance exists. It was pointed out at the conference that, because the resources available for enforcement have been limited, the enforcement program now depends a great deal on the voluntary efforts of the regulated community. The committee's view is that the enforcement program should move toward a policy of voluntary compliance rather than the employment of sufficient numbers of inspectors to totally police the industry.

Enforcement Versus Compliance

A review of the regulations indicates that some are written from a legalistic point of view. It is readily admitted that such regulations are written with the intent of making enforcement against violations of such regulations as successful as possible. Such a philosophy may stem from a view that at least some persons involved with the regulations will make little, if any, attempt to comply with them. Therefore, they must be written to be "violation proof." This idea seems to prevail even though enforcement of such sections is infrequent and the objective results in complex, hard-to-understand regulations.

The question may be raised whether it is appropriate to assume that shippers and carriers desire to comply with the safety regulations and, therefore, that the regulations should be written in a clear, concise, and uncomplicated manner. Such a change in philosophy might be characterized as enforcement versus compliance.

If it is true that the existing regulations are so complex that many people subject to them cannot understand them--and, as a result, there is noncompliance with the regulations--one way to achieve better compliance would be to simplify the regulations. In view of the allegations that there is substantial noncompliance now, simplified, more easily understood regulations could lead to substantially better compliance.

Preventive Enforcement

The principle of "pro-active" enforcement should also be considered. Pro-active enforcement is a term that some apply to inspections of vehicles and techniques before transportation occurs in order to inform and aid shippers, carriers, and transporters in regulatory compliance.

The various techniques for enforcement need to be governed by a developed concept that relates regulations to transport safety. Social regulation to obtain a desired goal such as safety is more complex than mere enforcement of regulations. It can involve both civil and criminal liability. Both regulation and the criminal law constitute the direct exercise of governmental power to coerce conduct perceived to be socially desirable or to prohibit or restrict conduct perceived to be socially undesirable. Historically, and for constitutional reasons, the operation of the criminal law system depends on the separate exercise of legislative, judicial, and executive powers. Regulation, however, concentrates power in a single, specialized body endowed with legislative authority to define the specific content of required or restricted conduct, executive authority to investigate and enforce compliance with regulatory standards, and, usually in connection with economic regulation, jurisdiction to adjudicate disputes between private parties.

Social regulation and, in particular, safety regulation relate primarily to the protection of certain persons such as employees, consumers, or motorists against risks created by other persons such as employers, manufacturers, or carriers. Thus, because the relation between risk makers and risk takers is noncontractual, it is improbable that alternatives to social regulation could be found in market mechanisms. If such alternatives exist, they are likely to be found in an exploration of well-established, though continually developing, civil law pertaining to noncontractual liability.

Civil Liability as Compensation for Private Injury

The law defines a tort as a civil wrong independent of contract. Although it is common to refer to the tort system as based on the "fault" of the responsible party, the commission of a tort may result not only from a willful or deliberate act or a knowing failure to act, but from an inadvertent or negligent failure to exercise reasonable care. The legal inquiry is whether the party charged with the commission of a tort acted or failed to act as a reasonably prudent person would have acted under all of the circumstances in a particular situation. In contrast to the consequences of criminal behavior, however, the mere commission of a tort creates no right to recover damages unless some harm or injury results, and then only when the tort is the "proximate cause" of such injury.

Because the development of regulatory mechanisms was, at least in part, a response to perceived failures of the tort system, some of these failures should be stated. The deficiencies frequently associated with the tort system include high cost, delay, proof of liability, identity of responsible parties, available defenses, and immunity from judgment.

It is generally agreed that "direct" damages recoverable under the tort system include only medical expenses and lost wages or income in the case of personal injuries, and the cost of repair or replacement in the case of damage to property. Customarily, however, the fault-based liability system also allows "incidental" damages, such as pain and suffering, which may far exceed the amount of the direct damages. The availability of such incidental damages may be at least partially responsible for inducing or prolonging litigation and, unfortunately, may provide an incentive for fraudulent or unjustified claims. More importantly, such damages may be so enormous that a single incident of disastrous proportions, involving multiple claimants, may deplete the assets of a sizable enterprise.

Alternative Systems of Liability

Strict liability, also referred to as liability without fault or absolute liability, permits recovery for "abnormally" dangerous or "ultrahazardous" activities even in the absence of fault or negligence.

Although the courts have generally restrained expansive application of the doctrine, strict liability has found increasing acceptance in legislative enactments. Thus, to one degree or another, a strict liability regime has been incorporated in such diverse legislation as state child labor laws, federal and state pure food laws, and railroad safety statutes (see, for example, the Federal Safety Appliance Act, 45 USC Sect. 1, et. seq.). The most recent federal adoption of such a regime is the so-called Superfund Act (CERCLA).

The so-called "compensation" system is best known in connection with employees' claims against employers for work-related injuries. Although sometimes referred to as an "insurance" system, it is a comprehensive legislative scheme characterized by (a) strict liability; (b) compulsory insurance; (c) administrative, as opposed to judicial, hearings on claims; (d) limited recovery for injuries, not including incidental damages for pain and suffering or similar injuries; and (e) limited attorney's fees.

Despite the model of the biblical Good Samaritan, physicians who happen to be present at the scene of an accident may sometimes be reluctant to volunteer expert advice or assistance. Such conduct may expose them to substantial liability if such advice or assistance is later found negligent.

In an effort to avoid such consequences and thereby encourage physicians to volunteer when needed, some states have enacted legislation immunizing physicians against liability under the medical good-samaritan laws. However, such legislation does not necessarily protect them against the cost, inconvenience, and professional embarrassment of litigation.

During the past two decades, a number of shippers of hazardous materials have developed emergency assistance programs whereby personnel employed by such shippers are made available as technical experts in connection with the on-scene disposition of a transportation emergency. Since a shipper is rarely liable for injuries that result from the operations of an independent carrier, the advice or assistance so provided exposes the good-samaritan shipper to liability that the shipper otherwise would not have. Therefore, it has been suggested that similar exculpatory protection should be provided in such cases. Apparently in response to such suggestions, the Superfund Act contains what appears to be the first good-samaritan provision under federal law.

Because most liability litigation is governed by state law, action by state legislatures would also seem to be necessary to relieve the good-samaritan shipper of potential liability.

Compulsory insurance schemes and government compensation funds demonstrate additional techniques that have been employed in an effort to ensure adequate compensation to injured parties.

Compulsory insurance is well known to the general public since many states now require liability insurance as a prerequisite to motor vehicle operation. The Motor Carrier Act of 1935 requires regulated motor carriers to obtain and submit insurance or other evidence of "financial responsibility" for payment of damages to injured parties. The Motor Carrier Act of 1980 extended these requirements to all interstate for-hire carriers and to all carriers of hazardous materials. Similarly, the Resource Conservation and Recovery Act of 1976, the Clean Water Act, and Superfund require certain persons to provide evidence of financial responsibility. [Ed. note: The minimum levels of financial responsibility requirements for carriers were superseded by Section 406 of the Surface Transportation Assistance Act of 1982 (96 Stat. 2097).]

An additional device for ensuring such recoveries (or payment for clean-up costs for environmental damage) is a government-administered liability fund. Such funds have been created under a variety of statutes, including the Deepwater Port Act of 1974, the Outer Continental Shelf Lands Act Amendments of 1978, and the Superfund.

Neither compulsory insurance schemes nor liability fund programs are limited to any particular systems of liability. For example, the Motor Carrier Act requires that insurance be provided by motor carriers, whose operations are normally subject to ordinary tort liability rules that do not limit the amount of recovery. At the same time, under various environmental laws, insurance is made mandatory in conjunction with strict liability and a limited dollar amount of recovery. Similarly, insurance is an essential feature of the compensation system, which combines strict liability and recoverable damages of a limited nature.

Civil Liability as an Instrument of Social (Safety) Policy

The system of civil liability, in addition to its function of providing redress for private injury, also serves as an important instrument of social policy because the likelihood of damages tends to restrain socially undesirable behavior. Both federal and state legislatures have tended to supplement the tort system with greatly expanded regulation and, in some cases, to modify or replace it with other mechanisms. Such legislative intervention, most of which has occurred during the past four decades, has been largely piecemeal, and little effort has been directed toward the establishment of a comprehensive and integrated liability system that is logically related to a consistent set of social objectives.

Liability from Noncompliance with Regulation

Among the pervasive consequences of such legislation, although apparently disregarded by many critics of the regulatory process, has been the expansion of the well-established principle that violation of a criminal or other statute that requires or proscribes specified behavior constitutes negligence per se and, therefore, subjects the violator to liability for civil damages. Although such statutes frequently require proof of criminal or specific intent, the principle has been extended to regulatory violations, though similar proof is rarely necessary.

One of the factors restraining broader application of the noncompliance principle has been the insistence by the courts on evidence that the violation was the proximate or probable cause of the plaintiff's injury.

Compliance as a Defense Against Liability

If noncompliance with regulation constitutes negligence, it would seem to follow that regulatory compliance should afford a defense in liability litigation. Nevertheless, with rare exceptions, the courts have concluded that mere compliance is not an absolute defense because the regulatory requirement may constitute only a minimum standard of safety or may be outdated. The validity of such reasons, however, may be questionable when, as in the case of many DOT packaging specifications, deviations from the required standard would be illegal even if such deviation proved to be safer than the standard itself.

It is sometimes argued that adherence to regulation should be deemed an absolute defense against liability in order to provide an incentive to compliance. Such a thesis, however, is not persuasive in view of the incentive furnished by the corollary rule attaching liability to noncompliance. Also, to permit the assertion of compliance as an ultimate defense might eventually prove even more distasteful than the denial of such an assertion. A likely result would be the generation of excessive pressures on regulatory bodies for further proliferation of increasingly detailed and stringent requirements. Finally, even if such an absolute defense rule might serve the interests of the public generally, there would be a substantial question of equity as to why any person who has suffered serious injury should be precluded from damages merely because those whose actions contributed to such injury complied with an outdated or otherwise insufficient regulation as a result of government neglect or misapprehension.

The problem of such government error, of course, might be resolved by the assumption of liability in such cases by the responsible government. Although the Federal Tort Claims Act waives federal immunity from liability in tort, it simultaneously prohibits government liability for an exercise of or failure to exercise a "discretionary function or duty." Thus, since the promulgation of regulations clearly requires the exercise of discretion, it appears that no liability would attach thereto absent an amendment of the Tort Claims Act--an unlikely prospect.

Need for New Approaches

To justify statutory limitations on recovering damages, it often is asserted that such limitations permit the insurability of otherwise prohibitive risks, thereby ensuring the viability of enterprises whose continued existence is considered essential to society. Nonetheless, it may be difficult to comprehend why particular individuals should be left to suffer the burden of uncompensated loss resulting from legislative limitations of liability while the benefits thereof accrue to others, along with society as a whole. Should not such losses more properly be borne in equal proportions by all who directly or indirectly enjoy such benefits? The same question, of course, may be raised in connection with exculpatory legislation, including good-samaritan laws, which are in effect the ultimate extension of the limited-liability concept.

The compensation system, while mitigating or eliminating many of the defects in the tort system, is simultaneously afflicted with the problems inherent in strict liability, limited recovery, and compulsory insurance systems, all of which are integral parts of the compensation scheme. Nevertheless, the compensation system seems to have enjoyed substantial approbation among a variety of interests and appears to reflect a series of practical and reasonably equitable trade-offs among the interests of all concerned. An additional attraction of that system is the incentive it provides to channel productive energies into the avoidance of injury instead of the tactics and strategy for winning lawsuits.

Notwithstanding the varied activities of legislative bodies, the courts have likewise searched for new ways to allocate the burden of damages. The direction of that search in recent years has been toward imposing a larger share of the burden on those with the greatest ability to pay. Accordingly, the interest of industry in its own survival compels serious consideration of alternatives to the tort system as society grows even more litigious and the judicial system finds novel ways to compensate the injured. One such alternative may lie in more extensive reliance on the compensation system to afford deserving claimants an expeditious method of fair recovery without the uncertain gamble of tort litigation.

It seems entirely possible that further exploration would reveal opportunities where a modified and, perhaps, more constructive liability system could be fairly and effectively substituted for a portion of government regulation, including some of the minutely detailed hazardous materials regulations.

It is not suggested here, however, that any liability system should adequately replace regulation where the transportation of hazardous products involves serious potential for catastrophe. It is possible that no existing or revised system of liability will prove fully satisfactory in sorting out the multiple possibilities of individual or joint liability associated with railroad accidents. In such complex situations, however, some reasonable combination of strict and tort liability could provide a creative matrix for limiting the extent of regulation or the necessity of protracted litigation.

At the opposite end of the spectrum, experience has demonstrated that transportation of many products shipped in less-than-bulk quantities creates little likelihood of serious harm. Although the transportation of paints and related materials, for example, produced thousands of reported "incidents" in recent years, such incidents have resulted in no fatalities and relatively few injuries. To maintain intensive regulation of such transportation, when a modified liability system might serve as an equal or better deterrent to unsafe behavior, tends to waste the resources of both government and industry, probably inhibits the development of improved safety methods, and detracts from the achievement of more important objectives of transportation safety regulation.

It is submitted, therefore, that regulation is not necessarily the exclusive mechanism for the achievement of reasonable social objectives. Alternatives to social regulation must be actively and vigorously pursued. The proliferation of regulation during the past two decades suggests that another century should not pass before such pursuit is begun.

Little use has been made of criminal sanctions in the enforcement of the laws that affect the transportation of hazardous materials, much less those related to the broader areas of health, safety, and environmental laws. To date, there have been no reported criminal convictions of corporate officers under the HMTA, the Consumer Product Safety Act of 1972, the Occupational Safety and Health Act of 1970, or the Toxic Substances Control Act of 1976. Until 1979, there were no reported criminal convictions under the Federal Water Pollution Control Act.

The only significant criminal proceedings in the transportation of hazardous materials are the indictments returned against five companies and one individual in connection with the November 1973 crash of a Pan American Boeing 707 at Logan Airport, Boston.

While there are many explanations for the lack of use of criminal sanctions against corporations and corporate executives in the hazardous materials area and the broader areas of safety, health, and environmental laws, one problem is related to pinpointing the blame for serious violations. Giant corporations with multiple layers of management have significantly complicated the critical process of fairly pinpointing such blame.

Yet, notwithstanding the difficulty of penetrating corporate management structures, the application of criminal sanctions has begun in the broad areas of safety, health, and environmental laws. Several cases have been referred to the U.S. Department of Justice for criminal prosecution under Section 17(e) of the Occupational Safety and Health Act. EPA expected as many as 50 prosecutions per year, beginning in fiscal year 1980, as the result of accelerated investigations by EPA and the Justice Department into hazardous waste dumping.

The Justice Department has implemented new priorities for investigating and processing white-collar crime. The proposed new Federal Criminal Code would add the business crimes of consumer fraud and a new felony called endangerment. Endangerment is a form of assault and would be present "where an individual's conduct manifests an extreme indifference or an unjustified disregard for human life." Endangerment would be associated with federal environmental, transportation, OSHA, and similar safety laws. Quite apart from this general approach, Congress has already adopted "knowing endangerment" as a crime as part of the Federal Solid Waste Disposal Act.

In punishing illicit corporate purpose, the primary rationale appears to be deterrence. Elements of retribution are also present because of the introduction of the elements of moral culpability in statutory and regulatory models. Such words as "willful," "knowingly," "with knowledge," and "intent to injure" reflect elements of moral culpability in that they require a criminal state of mind and a culpable mental state or mens rea.

In light of the substantive and procedural difficulties in the use of criminal sanctions and in consideration of a uniform, effective, and fair enforcement policy, it would appear that the preferred use of criminal sanctions in the enforcement of hazardous materials transportation regulations would be reserved for exceptional cases and not for day-to-day policing of the regulations. The latter can be left much more readily to aggressive inspection, monitoring, and enforcement with civil actions, injunctions, and, possibly, private treble damage actions of a nature now found under antitrust laws.

In using criminal sanctions in exceptional cases, concentration should be on deterrence and punishment of particularly recalcitrant or egregious corporate behavior. Primary targets of enforcement should be willful, intentional, or repeat offenders who violate important and substantive regulations. Prosecution of individual corporate officers should be attempted only where the evidence demonstrates that intentional corporate noncompliance with the law is the direct result of an informed policy decision made by such corporate officials. Further, criminal sanctions should be employed only when knowing or willful violations can be proved readily. Bringing a marginal criminal case in which the proof is weak, particularly against individual officers or employees of a corporation, will have little deterrent value if the agency is unsuccessful. On the other hand, it can have a devastating effect on careers and the general reputation of the individuals involved.

Finally, it is submitted that a comprehensive enforcement policy statement should be adopted and made public. The publication of such an enforcement policy would ensure that organizations and individuals subject to the provisions of the HMTA are aware of the types of violations under which the agency will seek criminal sanctions. Once such a rationale is made public, it has an essential element of fairness that will go far toward blunting any criticism of the agency's approach of preferring civil instead of criminal charges. It will eliminate the element of surprise, and will further a uniform, effective, and fair enforcement policy.

RECOMMENDATION 3: DEFINITION OF ROLES

In the discharge of its responsibilities as lead agency, DOT should meet with the body of industry and government representatives, convened pursuant to other recommendations of this conference committee, and should engage in role definition, emphasizing the value of uniformity of procedure and action. Congress should clarify the HMTA to remove the ambiguity caused by the term "inconsistent" in the preemption section of the law. In this process, Congress also should define the relative roles of DOT and the courts in determining disputes and, if that is to be DOT's responsibility, should equip that agency with the quasi-judicial tools to carry it out.

In addition, state and local governments should have a distinct role in the enforcement of hazardous materials regulation. Instead of truck inspections by the BMCS that focus primarily on shipping papers and placarding, DOT should also identify topics for enforcement that contribute more directly to accident reduction.

BMCS, as well as other modal agencies, should be reexamined. A consistent, cohesive, national enforcement program needs to be developed that uses the cooperation and efforts of industry and of state and local agencies. Guidelines for model state and federal efforts across all modes should be provided.

In such a national enforcement plan, concepts such as developing the use of "key state" enforcement efforts should be considered because of limited resources. Key states are those that have particularly heavy hazardous materials traffic. Also, pro-active enforcement to advise and educate shippers, carriers, and transporters of potential violations before vehicles are placed in transport should be evaluated.

DOT should investigate and develop guidelines for cost-effective, uniform state and industry enforcement programs. This advisory role is not meant to deliver DOT from any specific enforcement activities. The role of the states in enforcement needs to be encouraged in accordance with a master plan for national cohesiveness and consistency.

Regulation includes the enforcement of requirements. The writing of regulations and their enforcement must be directly related, with both functions geared to achieve the greatest safety for the greatest number of people. Through use of historical data and prospective analysis, DOT should advise all levels of government engaged in enforcement of the critical areas for particular emphasis. In addition, publication of a list of the particular provisions that are tied most closely to public safety can guide both the enforcers and the industry in the allocation of their attention. Enforcement of purely technical requirements is counterproductive; it demeans the program and distracts limited resources from areas where they may be more effectively used.

An affirmative effort should be undertaken to explore the implementation of criminal penalties, in circumstances where such prosecution would be appropriate in light of the committee's findings, and reliance on existing or legislated civil liability in lieu of or in addition to current enforcement programs.

FINDING 4: EMERGENCY RESPONSE

Given that no system of regulation and enforcement can eliminate accidents, a reasoned response to such occurrences is important. The Steering Committee found that there is considerable disagreement on the procedures and adequacy of emergency responses. However, there is agreement that accurate, timely, and continuous information flow is necessary at the scene of the accident in order to properly evaluate the hazard involved.

The response to an emergency that involves hazardous materials joins public and private emergency-response teams under stress conditions. These teams have differing purposes for responding. The public teams, i.e., fire and police personnel, etc., are there to protect public safety, health, and property. The transportation system teams, i.e., maintenance crews, track crews, etc., are there to restore the system as quickly and safely as possible. The private team, i.e., the hazardous material manufacturer, the shipper, etc., is there to provide assistance relative to the materials that may affect public health, the environment, and safety. With diverse and overlapping purposes, the need for planning, operational strategies, and designation of decisionmakers is very important. The reactions of the initial responders, most often the public agencies, are of the utmost importance. However, unless properly prepared, these people may be the least knowledgeable in the handling of hazardous materials incidents. Therefore, training programs designed for public emergency teams are necessary.

Communication channels should be opened to eliminate barriers between the responding teams. Methods of operation of response teams may need review and revision.

The committee arrived at this finding on the importance of emergency response after examining the issues of public and private roles, planning, response, communications, and emergency information needs. These are discussed below.

Public and Private Roles

There is a lack of understanding of what the appropriate public and private roles are. This provides the opportunity for errors, confusion, and perhaps even confrontation.

There is a tendency to consider these roles only in terms of an emergency situation and then only in terms of whose authority is final. Each group has a role long before the emergency occurs, and, if that role is properly understood and played, the question of authority in an emergency will become less important. The primary pre-emergency role is communication. Too often, public and private officials fail to develop lines of communication with each other. Lack of pre-emergency communication contributes to communication breakdowns during emergency incidents. There are many ways to improve communication that will produce useful results and will build the mutual trust needed at the scene.

Industry must encourage, and public officials must aggressively pursue, joint pre-emergency communications with the handlers of hazardous materials emergencies. Joint planning and training sessions are the logical beginning. They should include joint emergency planning for incidents, training exercises, and communication. Private industry must take a leadership role through support of the education and training of public safety personnel in the communities in which they do business or through which their products may pass.

Shippers have a definite role to play during a hazardous materials emergency. It represents a resource that has been under-used for many reasons, not the least of which is the attitude of fire departments toward their contribution. Through day-to-day involvement with hazardous materials, private industry is intimately familiar with specific hazards, precautions, procedures, equipment, and materials associated with each chemical. They will often have an inventory of special materials and equipment required in developing, processing, storing, or transporting the chemical. They also have personnel who have knowledge of the hazards and precautions involved and are trained in the use of special equipment and materials.

Because the private sector can be of vital assistance in emergency response, it is important that an atmosphere be created in which industry is willing to assist. Good-samaritan laws to protect qualified industry personnel from liability are to be encouraged. However, it is also important that such laws distinguish between the unqualified and those who have demonstrated competence.

Industry's involvement with hazardous materials usually occurs under the controlled conditions of the normal work environment and not that of an emergency scene. Although larger companies may have an emergency-response team, most do not. Those that do not should not be expected to perform tasks that differ from their normal job under the pressures of an emergency. During the emergency, their role should be one of advice and support. When conditions have been stabilized and the stresses relieved, they should be expected to perform tasks that may be associated with product control, transfer, or cleanup and should do so in cooperation with the public agency in charge.

In the area of emergency response, some agencies play only a supportive and advisory role. It is important, then, that one public agency be given the responsibility and authority to assume command of operations at the scene of the incident. The agency will vary from jurisdiction to jurisdiction. The most important thing is that one agency be clearly designated before the incident, and that this designation be clearly communicated to all parties involved. In states or cities that have adopted the Uniform Fire Code, the command agency is the fire department. This code assigns to the fire chief many of the responsibilities that are commonly associated with hazardous materials incidents.

Planning

One goal of planning is to increase effectiveness during an emergency. Potential problems and their possible effects must be anticipated and solutions developed prior to their occurrence.

Most hazardous materials incidents are of a minor

nature. They can be handled by the people and equipment at hand. But even a minor incident may become a major incident and that major incident can tax resources to the limit, or beyond, especially if proper planning has not been done. There are numerous examples of bad decisions based on inaccurate or inadequate information--information that planning could have provided.

In order to prepare plans for a hazardous materials incident, public safety officials should know the materials commonly passing through their jurisdiction. As noted earlier in this report, such information is not usually available.

Better information is needed on the behavior of hazardous materials under emergency conditions in order to anticipate problems. In many cases, these data are available, as is the technology for delivering the data. What is lacking is an appreciation of the usefulness of the information.

Those responding to an emergency must have information immediately available on how the involved materials may react--not under normal or laboratory conditions but under emergency conditions. Emergency-response data should tell what actions can be taken to protect large numbers of people when evacuation is not possible.

Response

In response to any hazardous materials emergency, there are a few basic principles that must be kept clearly in mind. The primary objective is to minimize damage and injury. While time may be an important factor, it usually is not the paramount one.

A serious hazardous materials incident will probably be too complex for any single individual or organization to have all the information and answers. To cope with such incidents takes emergency management, and a management-team approach has the best chance of success. Unfortunately, there are several factors that may make such an approach difficult. One is a lack of trust and understanding between the various groups involved. Another is fear of liability for what may occur during the emergency.

A third factor essential in successful emergency management is unity of command--i.e., one person must be in charge and must accept responsibility for what occurs. Command must be given to that person who has the legal responsibility for the public safety of the citizens in that area. In any state that has adopted the Uniform Fire Code, that person is the fire chief.

Aside from legal responsibilities, there are other reasons why the fire chief is the logical choice. The fire department often is the first emergency-response team on the scene. The fire department will have to deal with the emergency for some time before shipper representatives or government agencies arrive, and there is an excellent chance that many critical decisions will be made prior to the arrival of this assistance.

The fire chief may not be the most knowledgeable person present on the nature of the material involved and on the technical procedures needed to deal with the problem. Thus, fire chiefs or others designated to be in charge must seek all advice and counsel available to them. They must be trained to establish a command post, to gather advisors, to weigh and consider advice, and to act on it. At the same time, advisors must realize that the commanding officer has the responsibility for public safety and for the actions to be taken.

Communication

On-scene communication, both face-to-face and over distance, is vital. Members of the teams must maintain contact with each other and with those in charge at the command post. Since transmissions over distance rely heavily on radio, it is important that planning for emergency management consider the transmitting and receiving capabilities of each team. The command post must be equipped with modern equipment to facilitate communication over distance.

Another area of communication is with the public, including the release of information to the media and giving public warning. Press and media relations are important parts of emergency scene management. The public has a right to know what is happening in their community, and accurate and responsible reporting of an incident can meet that need. Conversely, a lack of accurate information can lead to very undesirable consequences.

A major hazardous materials incident in a metropolitan area may necessitate warning thousands of people. Evacuation will be difficult, perhaps impractical, or even impossible. In spite of this, in some situations people must be warned to take some action to avoid injury. There is no practical way of giving such public warning. Broadcast media will reach those who happen to be listening or watching. Police may be able to warn those in the streets. Despite the best efforts, however, many will not hear the warning.

One answer might lie in the installation of warning devices in major areas of occupancy that could be activated by authorities. Whatever the answer, the whole issue of public warning and avoidance measures is one that needs serious consideration and planning.

Emergency Information

Those managing an incident need accurate, timely information. The information base is so complex that a concerted effort is required to provide an adequate resource file at the scene. The file must contain information in a format that is quickly understandable to people under the stress of an emergency.

The cooperative efforts of CHEMTREC and the National Response Center to provide this type of information are recognized as valuable and deserving of encouragement and support. A systematic research effort to develop a reliable data base for use at emergency scenes needs to be sustained because of the dynamic nature of hazardous materials transport. Such data base needs are not satisfied by safety data sheets or emergency-response guidebooks alone. These are useful efforts that should be continued, improved, and refined, but they do not provide sufficient information for many circumstances.

Shipping papers, placarding with material identification numbers, and labeling also contribute valuable information. They fail, however, unless they are uniformly enforced. If an emergency-response team cannot have confidence that a vehicle is properly placarded, then the vehicle might as well not be placarded at all. Those at the scene must assume that the placarding information presented is correct.

In the vast majority of cases, the response to the scene of a hazardous materials incident will not involve federal response personnel. Thus, the target for response information at the scene must be local personnel. The systems developed to meet on-scene information needs must be designed for the local responders, who have the most frequent need for such data.

RECOMMENDATION 4: DEVELOPMENT OF INFORMATION

Federal agencies with emergency-response duties should take action to develop the required information and technology to provide accurate and fast identification and estimation (both remote and on scene) of the nature and degree of hazard in hazardous materials emergencies. Congress should support research to improve evaluation of on-scene hazards and to facilitate communication between all responsible individuals and teams as well as with the general public.

A great deal of effort must go into emergency response. Much disagreement exists on what is needed. The conference committee recommends that immediate attention be given to the three areas of planning, training, and designation of who is in charge of hazardous materials transportation emergencies. Whereas in the past the federal role has been directed almost exclusively to the pre-accident actions of industry, the post-accident actions of all parties demand thorough examination at the federal level.

FINDING 5: TRAINING

There is no systematic federal approach to training in the area of hazardous materials transportation.

In a follow-up to the ten critical issues survey conducted among the invitees to the Williamsburg conference, the issue most frequently included was the need for training for the more than 2 million persons involved in all aspects of hazardous materials transportation. Such groups include the following:

1. Personnel involved in establishing regulations and enforcing them;
2. Shippers whose responsibility involves the classification, packaging, marking, labeling, and documentation of the hazardous materials to be transported;
3. Carrier personnel who have the responsibility for transporting hazardous materials in commerce;
4. Personnel who receive the materials from the carriers and must handle them in their dissemination to users; and
5. Emergency-response personnel who respond to an incident involving hazardous materials.

All of these must be trained in the proper performance of their jobs and in the regulations as they affect these jobs.

Noncompliance with regulations that govern hazardous materials transportation has been attributed by many federal agencies to a lack of knowledge and training among both carriers and shippers. Similarly, the training of response personnel has been inadequate.

Basically, all training for hazardous materials can be divided into three major categories: prevention, initial response, and reflective response.

Prevention

Prevention refers to those actions of industry and regulatory authorities to ensure that the product is properly classified, packaged, labeled, documented, and safely transported.

DOT, through the MTB and the four modal administrations, writes and enforces these regulations. Their purpose is to package the product in such a way that it will be safely contained during the expected normal rigors of transportation. The federal agencies are active in enforcing these regulations.

The training of inspectors consists of familiarization with the regulations followed by on-the-job training and experience. Inspectors can be effective immediately due to the fairly low risk involved and the fact that questions can be directed to superiors or to the MTB. It is also important to note that, besides the federal agencies, some states take an effective preventive role.

Transportation incidents that involve hazardous materials occur in proportion to the number of vehicles carrying hazardous materials compared with the total number of vehicles. In other words, the presence of hazardous materials neither adds to nor detracts from the possibility of a transportation incident.

Initial Response

Initial response refers to actions taken immediately after a hazardous materials accident or incident by industry and the various local, state, and federal response agencies, seeking to minimize and control the effects of the accident or incident.

If a transportation incident involving hazardous materials occurs, responding personnel are most often, if not always, those who would respond to any transportation incident. Their actions may be extremely critical. An incident improperly handled can become a catastrophe.

Unfortunately, first-response personnel are likely to have received minimal training in proper response to a hazardous materials incident. Technical qualifications of response personnel in larger cities usually are better than those of volunteer units, but the technical competence and hazardous materials background of the majority of police, emergency medical personnel, and firefighters are generally less than adequate. Few localities are aware of the type and quantity of hazardous materials that move by rail or highway through their area. Most are also unaware of the implications of the various physical, chemical, and toxicological properties of hazardous materials.

It is important to note here that the vast majority of initial-response personnel will never be involved in an incident in which a hazardous material is released. Also, most initial-response personnel do not have the experience, technical background, or time to prepare for the potential ramifications of a hazardous materials incident.

Training in initial emergency response is probably the least available type of hazardous materials training. It is usually not offered at the location where it will be used, nor are the persons who will be involved in its use those who attend the courses.

Since the people who must be trained rarely use the training, they cannot be expected to apply it properly in a real situation. The initial response will be made by personnel who will be poorly prepared and inexperienced in responding to hazardous materials incidents.

Yet, the record is not unduly frightening. Hazardous materials incidents do occur and, through great effort, courage, and quick thinking, are fairly well handled by those who first respond. Initial response usually is based on the anticipated worst-case consequences, even though this does not always come about. Thus, the action taken is assumed to have been the appropriate one.

Personal risk to the initial-response forces can be extremely high. Even if the hazardous material can be readily identified, the proper types of respirators, protective clothing, and detection equipment are not always available.

Reflective Response

Reflective response is the response by knowledgeable and experienced personnel, usually after the initial-response forces have reported to the scene. The response is "reflective" in that the personnel have been made aware of the situation at the scene and have been given the time to consider the situation and act with some degree of confidence. Personnel in this category include chemical industry personnel, clean-up contractors, carrier representatives, federal on-scene coordinators, and their forces. Most of these personnel are experienced in responding to hazardous materials incidents. They have proper respiratory protection, contact protection, and detection equipment. A primary part of their job is to be prepared for such incidents, and they can call on established consultative services for advice and counsel.

Once reflective-response forces have reached the scene of a hazardous materials incident, they should be expected to direct or give advice to the initial-response forces.

The training of the reflective-response forces is through experience, involvement in past hazardous materials incidents, and technical training in scientific disciplines such as chemistry, physics, and engineering. There is tremendous interest in providing qualified, experienced reflective-response personnel. The shipper gains credibility and the carrier controls his own property and equipment. The level of training of such persons is considered adequate.

Training Deficiencies

Those most in need of training are the initial emergency-response forces. Such persons consist of state and local police and firefighters. To a lesser extent, emergency medical teams, public works crews, and personnel involved in disaster control and emergency preparedness can be included.

The factors related to this problem were identified by the conference committee as follows:

1. There is no direct coordination for initial-response forces at the local level.
2. Initial-response forces show a reluctance to accept coordination.
3. The technical ability, education, experience, supervision, equipment, training abilities, etc., of local forces vary widely.
4. There are difficulties in offering training to initial-response forces, who may have a lack of aptitude or may not be available for training at the time it is offered. These problems can be particularly severe for volunteer groups.
5. There is no mandated training requirement for initial-response forces.
6. There are often difficulties at the scene of an accident with the jurisdictional scope and priority of various groups.
7. Vehicle operators should be better prepared to meet their obligations under the law in an accident.
8. Standardization of contingency planning is required.
9. There is a lack of qualified instructors, funds, etc., for training emergency-response forces.
10. There is a lack of knowledge about the availability of emergency-response courses at the state and local levels.

Training Program Development

The training of persons involved in hazardous materials transportation is complicated by the lack of nationwide minimum training criteria and the absence of a coordinating agency with authority to develop standards and plan and implement training programs.

Factors that affect the development of uniform training programs include the following:

1. The size and diversity of the audience and the difficulty of identifying this audience;
2. The complexity, scope, and changing nature of the regulations governing hazardous materials transportation and the level of understanding and acceptance of these regulations by all parties;
3. Management and employee perspectives on the need for training;
4. Funding;
5. Inconsistencies in training programs and the lack of qualified instructors;
6. The diversity of the training programs required;
7. The need for ongoing and specialized training; and
8. The need to tailor the training to fit job functions.

Training is planned learning activity designed to bring about changes in the behavior of the learner. In other words, we train people to do what we think they should do in given situations. Typically, we want personnel to perform defined tasks under specific conditions and in certain circumstances, so training directors develop detailed projections of future situations and the sets of respondent behavior that experts believe are most desirable. Training programs should be based to a large extent on experts' predictions of future situations and problems, so that students will be trained to bring incidents to a satisfactory resolution.

The ability of the experts to predict future problems depends on their ability to understand history, to analyze technical data, and to generate insights. Those who invest in training want to be sure that future situations will be met by "approved" behavior on the part of the trainees. Experience with hazardous materials training points out the importance of the following concerns:

1. Training goals, or accurate prediction of future problems;
2. Reality-based curriculum;
3. Methods of teaching;
4. Performance-based instruction;
5. Standardized instruction;
6. Certified instructors;
7. Certified or licensed graduates; and
8. In-service training.

Training should be a total system. The pressure for quick results often forces the operation of only a part of the training system, thus reducing the chance for long-term success.

Training Responsibility

The HMTA designates DOT as the agency to establish and maintain a central reporting system and data center capable of furnishing technical advice to law enforcement and firefighting personnel that will aid them in responding to emergencies that arise from transporting hazardous materials. The Act also specifies that advanced training of emergency-response personnel should be considered an integral part of an adequate response system. DOT is further authorized to establish criteria for handling hazardous materials. Among such criteria cited in the HMTA are personnel training and qualification.

The Williamsburg conferees recognize the designation of DOT as the lead agency for development of a master program for training associated with transportation of hazardous materials for all personnel involved in public and private organizations and for all phases of hazardous materials transport.

DOT should, as appropriate with other relevant federal agencies, establish a coordinated advisory group, or task force, made up of representatives from government, emergency services, and industries concerned with the transportation of hazardous materials.

This task force would be authorized to (a) be responsible for development of policy regarding training, (b) identify the groups that need training and the skills needed by each group, (c) promote recognition of the need for proper training of all groups, (d) establish criteria for training curricula, (e) examine the need for regional training networks, and (f) establish a priority scheme for training that considers the urgency of the problem in various areas of the transportation system.

Funding for training would be the responsibility of the coordinating agency. It is further recommended that DOT, acting as the national coordinating agency, establish an internal, centralized office to develop for the states, as an element of a national contingency plan, a model program for initial-response training. Priority would be given to the following:

1. Identification and evaluation of the present training courses in emergency response and publication of lists of these courses;
2. Promotion of development in each state of an agency to direct and assess the initial-response force capability and its performance and to keep records on the personnel trained for initial response;
3. Work with state, police, and fire service organizations to serve as a training delivery system for courses in emergency response; and
4. Provision of courses at the federal level to train and certify instructors of initial-response courses at the state and local levels.

RECOMMENDATION 5: TRAINING PROGRAM

DOT should designate an office to develop a master program for hazardous materials training for personnel in public and private organizations and for all phases of the transportation of hazardous materials.

Instructor training and certification should be a major priority of this office. Industry, emergency groups, and others must be involved in the formulation of this master program. The office would be responsible for distribution of available funds for training and would identify and evaluate current federal, state, and industry programs.

FINDING 6: TECHNOLOGICAL INNOVATION

The committee found that there is a need for more technological innovation, information, and research in the transportation of hazardous materials. It also recognized, however, that existing technology and information are often not sufficiently used.

Overly specific regulations and lack of funds inhibit technological innovation, information, and research in the transportation of hazardous materials.

Background

The U.S. government has had a substantial influence on technical research and development (R&D) activity since World War II, supporting more than 50 percent of the nation's R&D investment for most of that period (18).

Although its direct involvement has been concentrated in the defense and health sectors, the government has influenced research in all segments of the economy, including the transportation of hazardous materials and wastes. Public R&D programs have generally served the purposes of either developing new technology for public-sector needs or advancing basic knowledge or understanding. For the most part, the federal government has avoided the support or conduct of research to develop new private-market products or services (25). Even so, the overall role of the federal government in supporting public technological R&D has been questioned in light of allegations of waste and mismanagement in some research programs.

The argument for reduced government involvement in R&D is based on the premise that government will generally be less efficient than private industry in directing R&D activities.

On the other hand, there appears to be a near consensus among economic and business analysts that the national investment in R&D needs to be increased from current levels if future gains in productivity and the standard of living are to be ensured. Given some uncertainty over the private market's willingness to significantly increase R&D investment, especially in areas such as hazardous materials safety, the federal government may be the only significant source for much of the needed additional research funds.

Although the U.S. private economy has had spectacular success in developing and bringing to the market a wide variety of commercial products, there are strong theoretical economic arguments that the private market has and will continue to fund R&D at below socially desirable levels. The most prominent reasons advanced to explain this underfunding of R&D include (a) lack of private-market economic incentive, (b) uncertainty of payoff from R&D investments, and (c) restrictive regulations.

The private economy has a natural incentive to invest in the generation of goods that produce business profit. However, goods such as safety and environmental protection, while valued highly by the public, cannot be owned and sold by firms that contribute to their production. Accordingly, private investment in these areas will generally be less than the socially desirable amount. In particular, private investments in the production of new technology or other means of reducing the consequences of hazardous materials spills will be made only to the extent that they are cost-effective in reducing liability and other private costs of accidents. Government has the justification and responsibility for intervening in the private market to influence the production of these public goods in adequate quantities.

The uncertainty of future costs and benefits of technological research is another reason cited for the private-market failure to provide an optimal allocation of resources to R&D. "The outcome of any research project is necessarily uncertain and the most important results are likely to come from projects whose degree of uncertainty to begin with was the greatest" (2). Because private firms have been found to be generally adverse to risk with respect to investment in R&D, they will tend to underinvest in technological research and skew their R&D investments away from basic or long-term research and toward applied, short-term endeavors. Because R&D expenditures in risky research are very likely to produce the most important benefits from society's point of view, an argument can be made for government intervention in the private R&D market, particularly in support of basic research.

Such arguments indicate that the specific areas in which the government could intervene in R&D to increase the general public welfare are difficult to define. Classical welfare economic theory gives little assistance. Its prescription, i.e., invest until the marginal social benefit just equals the marginal social cost, cannot be used in practice because of the uncertainty of estimates of social costs and benefits. Government intervention in the R&D process generally results in increased administrative costs and can lead to misdirection of private as well as public resources. Before specific public intervention can be justified, it is necessary to compare each option's prospects for remedying the market defects with the mischief that these options may themselves generate (44). The government, therefore, should be very careful in devising strategic and tactical plans for intervention in the technological R&D process. As a general rule, it should only intervene in areas where there is a clear societal benefit (measured by using marginal cost/benefit analysis in a qualitative manner, if necessary) and favor methods of intervention that cause the least disruption to the economic process.

Hazardous Materials Transportation R&D

A critical need for technological innovation arises from a pressing need for a solution to important problems. In the safe transportation of hazardous materials and wastes, several factors combine to lessen the critical nature of needs for technological innovation. First, the past safety record in hazardous materials transportation has been excellent, despite the media emphasis on accidents and the public perception of problems. Because of this, the benefit of R&D often becomes clouded; handy and popular solutions often go looking for problems to solve. A wide range of safety improvements could be implemented that require only political decisions, not technological developments. The implication is not that there will be insignificant payoff from application of technological developments in hazardous materials transportation, but that the areas where technological R&D investments should be made may be difficult to identify.

In discussing the government's role in developing technology and innovative techniques and hardware, the conference Workshop on Technology Development and Innovation concluded that government should avoid competition with private industry in those areas where the private sector is willing and capable of performing research and development. The workshop identified nine areas in which it was felt that technological innovation is critical. These areas are briefly noted below:

1. Performance Standards-Based Regulations--Engineered design standards form the basis for many federal regulations and may inhibit innovative hardware design and development of new technology contributory to the safe transportation of hazardous materials. In addition, design specifications are quickly outmoded. The use of performance standards in government regulations would encourage industry to develop, preferably through industrial consensus standards organizations, the best technological ways to meet the standards. Such a recommendation was made in the 1969 Airlie House report, but DOT has not fully implemented such procedures (17).

2. Technological Data Base--There is a need to pool all known technical data from both national and international sources to assess what is known, as well as what is lacking, in the development of safety standards. Such an assessment is necessary to evaluate the efficacy of performance standards.

3. Modification of Hazardous Materials--Lowering the hazards posed by materials in transport by modification of the physical and chemical forms of the materials and, through packaging, to achieve safety represent an area requiring more research. The use of inhibitors, suppressants, neutralization, gelation, expanded metal mesh containment, and improved package design need to be studied from the perspective of preparing materials for transport, transportation itself, and in the event of its release during an accident.

4. Evaluation of On-Scene Hazards--Information available at the scene of a hazardous materials accident is critical to decisionmaking. Information needed includes the state of integrity of the containment and a rapid and clear identification of the materials involved.

5. Cargo Tank Safety Devices--The safety devices on cargo tanks pose regulatory and design problems. Design improvements are needed for cargo tank piping, remote controls, fusible links, control cables, and other emergency systems. Greater reliability of these devices is also critical. More consistent and thorough enforcement to assure that compliance in the use and maintenance of safety devices is being achieved is needed.

6. Hazard Classification Criteria--Definitions, tests, and classification of hazardous materials are usually related to the in-plant use of the materials. Present safety regulatory criteria and schemes for classifying hazardous materials and wastes need to be examined for the adequacy of their relationship to the transportation environment. In addition, there is inconsistency in the definitions and classification protocols used by the different federal regulatory agencies.

7. Protective Clothing and Personal Equipment--Emergency-response and cleanup teams need protective, full-body suits that are lightweight, comfortable, and that permit long working hours, allow full mobility and dexterity, and assure communications. Customized protective clothing to be worn in hazardous materials accident response and cleanup needs to be more effectively designed and priced lower. Breathing apparatus that fits properly and allows work over a sustained period is also necessary.

8. Cargo Tank Truck Stability--The on-road stability of cargo tank trucks is an issue in the truck transport of hazardous materials. Vehicle rollover is a major source of leakage. Regulatory vehicle performance standards that address the problem of the total vehicle design, as opposed to just the tank design, are necessary. The impacts of such vehicles on highway design and regulations should also be examined as should the cost-risk benefit of design changes.

9. Pipelines--Pipelines, currently used primarily for the transport of petroleum products and some anhydrous ammonia, may have some potential for the transportation of other hazardous materials and wastes. The technological, safety, economic, and eminent domain issues surrounding the potential use

of pipelines for the transport of hazardous materials and wastes require exploration.

RECOMMENDATION 6: INNOVATIONS

DOT should continue to address and press for implementation of technological innovations that might improve hazardous materials transportation. The following areas for innovation should be considered:

1. Performance Standards-Based Regulations--Federal agencies should be encouraged, through Congressional action if necessary, to base new regulations (and to rewrite old regulations to do so) on performance standards. The standards should be practically designed to meet the need for safe transport of hazardous materials and wastes. The use of cost-risk-benefit analysis to determine where levels of safety performance should be changed in such a way as to stimulate industrial innovation is recommended.

2. Technological Data Base--DOT's RSPA should take the lead in reviewing the technological data base required for the establishment of performance standards, to identify gaps and needs, and to expand the data base as necessary. Where data are inadequate, performance standard development should start immediately.

3. Modification of Hazardous Materials--Federal agencies (both regulatory and non-regulatory) and industry should support the development of techniques and methodologies for rendering hazardous materials less hazardous in transport, in accidents, and at the accident scene during recovery and cleanup.

4. Evaluation of On-Scene Hazards--Federal regulatory agencies with emergency-response responsibilities [FEMA, DOT, DOE, EPA, NRC, and the Department of Defense (DOD)] should take the necessary actions to develop the required information and technology to provide for accurate and fast identification and estimation (both remote and on-scene) of the nature and degree of hazards that result from accidents and spills.

5. Cargo Tank Safety Devices--Government (both federal and state) regulatory agencies should determine the size and extent of the problem. Federal regulatory agencies should prescribe performance standards that encourage development of new safety device designs and should remove present regulatory inhibitions against technology development and innovative safety device design.

6. Hazard Classification Criteria--DOT's MTB should reexamine the present DOT-EPA definitions and classification protocols for clarity, applicability, currency, relevance to transport (normal and accident conditions), and interagency consistency. DOT's RSPA should, with industry input, develop whatever new technology is necessary to improve the transport hazard classification schemes. The Office of Management and Budget should require the federal regulatory agencies to be consistent in their definitions and classification protocols.

7. Protective Clothing and Personal Equipment--DOT, EPA, and the industry should work closely together to bring about better and more effective designs for improved personal protective clothing and equipment. The expertise and experience of the National Aeronautics and Space Administration and DOD should be used.

8. Cargo Tank Truck Stability--FHWA should prescribe regulatory performance standards for tank vehicle stability (especially for rollover and jackknife protection). It should remove any present regulatory inhibitions against technology development and innovative vehicle designs.

9. Pipelines--Industry (e.g., the Chemical Manufacturers Association and the American Petroleum Institute) should examine the need for this type of transport and should determine what technological development and innovations might be necessary to meet any identified needs. If the system proves feasible, Congress should act to facilitate the implementation of the technology, especially in the area of eminent domain legislation.

FINDING 7: DOT RESPONSIVENESS

Although encouraged by DOT's adoption of several of the recommendations from the 1969 Airlie House conference, the conference committee also noted that some of the most far-reaching proposals made there, e.g., performance standards, have not been adopted.

The Williamsburg conferees identified two recommendations from the 1969 Airlie House conference that have not received sufficient DOT attention. The first was the need to establish effective criteria for the classification of hazardous materials. At the time of the Williamsburg conference, problems still persisting included the definition and classification of flammable liquids and solids, corrosive solids, and hazardous wastes. The combination of hazards for a given material can include such factors as flammability, toxicity, and reactivity. Any or all of these factors contribute to the severity of the hazard presented by a material. Definitions and tests to identify the degree of hazard intrinsic to materials are usually related to the in-plant use of materials. A realistic assessment of the degree of hazard in the transportation of a material should be inextricably linked to the transport environment. A careful description, a classification scheme, and a procedure need to be developed for incorporating transport environmental characteristics with the characteristics of specific materials. Although some work has been accomplished on this subject, progress is slow.

Various reasons may be given for this slow progress. Not the least of these may be the lack of incentives in government and industry for greater activity and research. If it were not possible for industry to place a material in transport unless its behavior in the transport environment were known to be within the limits of clearly defined criteria, then adequate materials testing and environmental characteristics identification would be forthcoming. In the present environment of deregulation, this restriction on industry may not be well received.

A more aggressive effort from DOT through grants to academia and industry could be used to develop the needed structure, procedure, and criteria. Funds dedicated to this effort should be provided.

The second recommendation identified as not receiving appropriate attention is that a performance standard approach to regulation be used. The prevalence of engineering specifications adds unnecessary volume to the regulations and has inhibited creativity and problem solving. The greatest objection is that the use of performance standards complicates enforcement and the prosecution of violators. There are obvious trade-offs to be considered, but a concerted effort to use performance standards where appropriate is needed.

RECOMMENDATION 7: PERFORMANCE STANDARDS

Regulations should be improved by incorporating performance standards wherever and however it is feasible to do so. [Ed. note: Docket HM-181, Performance-Oriented Packaging Standards, announced April 15, 1982, is a move in this direction.]

DOT should lead a review of the technological data base required for the establishment of performance standards, identify needs, and expand the base as necessary. Where data are adequate, development of performance standards should begin immediately.

Appropriate industry and trade associations working with government should develop accepted-practice guides in lay terminology that conform to user needs in order to increase the understanding and effectiveness of regulations. Private-sector groups should take the lead in initiating this activity.

It is the committee's firm view that regulations governing hazardous materials transportation can and should be written simply, briefly, and effectively so that those subject to the regulations can understand them and all parties can have confidence in the level of safety achieved by compliance.

FINDING 8: PUBLIC PERCEPTIONS

The lack of safety in the transportation of hazardous materials is often exaggerated to the public.

Negative events often make news. Deviant behavior, natural and person-made disasters, conflicts, accidents, crimes, and delinquencies attract the news media. This fascination with the negative can result in distorted public perceptions of reality. Reporting of accidents that involve the transportation of hazardous materials is to be expected. There is a failure to report the overall excellence of safety in their transport and the intrinsic value of these materials to the nation's economy.

Earlier in this report, the essential description of the concern about hazardous materials was for a low-probability occurrence with a potential for high-severity impact. The public is made aware of the latter but not of the former. While responsible reporting should provide a reasonable perspective, it probably is not realistic to expect the media to fulfill this role. It is incumbent on industry and government to communicate the risks and benefits, including the relative safety, of hazardous materials transportation.

The flat opposition to the transportation, storage, disposal, or treatment of hazardous materials and wastes by members of some communities is, in the committee's view, an indication of the extent of the impact of the reporting of accidents and failures without compensatory exposure of the facts concerning the relative safety and essential benefits of these materials. Pervasive "anywhere-but-here" attitudes may lead to a disruption in the flow of essential materials through the promulgation of arbitrary regulations and the imposition of unnecessary bans. Mechanisms for conveying accurate information to the public regarding the issues inherent in the transportation of hazardous materials, as well as a means of conveying the public's concerns to the leaders of government and industry, are necessary and vital to the continuing and increasing movement of these products in commerce.

RECOMMENDATION 8: COMMUNICATION WITH THE PUBLIC

There should be a concerted effort by government and industry officials to communicate to the public the facts about the safety of hazardous materials transportation.

To the extent practicable, both industry and government should engage in a two-way dialog with the general public on the issues. This will ensure that correct and appropriate information as to the nature, risk, and safety of hazardous materials transport is conveyed to the public, and that the public's views are adequately reflected in public policy actions.

Leaders at all levels of government should develop and present to their constituent groups information that will enable such groups to develop an accurate perspective regarding the transportation of hazardous materials and wastes. DOT should develop public service messages that can be presented through the mass media. Trade associations should do likewise.

PART 5
Appendixes

APPENDIX I: References

1. H.S. Ang and J. Briscoe. Development of a Systems Risk Methodology for Single and Multi-Modal Transportation. U.S. Department of Transportation, Rept. DOT-RSPA-DPB-50-79-16, July 1979.
2. K. Arrow. Essays in the Theory of Risk-Bearing. Markham Publishing Co., Chicago, 1971.
3. Association of American Railroads. 1981 Annual Report. AAR, Washington, DC, 1981, p. 23.
4. D.M. Baldwin. Regulations of the Movement of Hazardous Cargos. National Cooperative Highway Research Program, Draft Final Rept., May 1980.
5. H. Brumm and J. Hemphill. Role of Government in the Allocation of Resources to Technological Innovation. National Science Foundation, Washington, DC, Rept. NSF/RAD-75/1/2, Feb. 1976.
6. Bureau of Explosives, Association of American Railroads. Emergency Handling of Hazardous Materials in Surface Transportation. AAR, Washington, DC, 1981.
7. W.J. Burns. Emergency Response. Presented at 59th Annual Meeting, TRB, Jan. 1980.
8. J.R. Cashman. On Site with a Hazardous Materials Team. Emergency, Vol. 11, No. 8, Aug. 1979, pp. 53-57.
9. P.C. Ditzel. Death of the Docks. Firehouse Magazine, Vol. 5, No. 2, Feb. 1980, pp. 46ff.
10. W.R. Fiste. Hazardous Materials and Highway Transportation. Presented at Annual Meeting of National Association of Governors Highway Safety Representatives, Stateline, NV, Oct. 25, 1978.
11. T.S. Glickman. Aggregate Effects of Rerouting Hazardous Materials Railroad Shipments. Research and Special Programs Administration, Transportation Systems Center, U.S. Department of Transportation, Cambridge, MA, Rept. PM-223-6-611-5, March 1982.
12. G.D. Hilberts, J.H. Berkley, Jr., and F.M. Quinn. Mississauga: Lessons in Large-Scale Evacuations. Proc., Hazardous Material Spills Conference, Milwaukee, WI, April 19-22, 1982, pp. 56-63.
13. Interindustry Task Force on Rail Transportation of Hazardous Materials. Systems Safety Analysis Subcommittee Phase I Final Report. Research and Test Department, Association of American Railroads, Chicago, Rept. R-344, Dec. 1978.
14. K.J. Kooyoomjian, B.A. Hostage, F.C. Gunderloy, G.R. Ricci, and R.S. Smith. Latest Developments on Hazardous Substances Designation and Multimedia Reportable Quantity Assignments Under Superfund. Presented at Hazardous Material Spills Conference, Milwaukee, WI, April 19-22, 1982.
15. Frank Malone. Commodities by Rail: Chemicals. Railway Age, Aug. 25, 1980, pp. 24-34.
16. National Academy of Engineering, Committee on Transportation. A Review of the Department of Transportation Research and Special Programs Administration's Hazardous Materials Research and Development Program. National Academy of Sciences, Washington, DC, 1980.
17. National Research Council. A Study of Transportation of Hazardous Materials; A Report to the Office of Hazardous Materials of the U.S. Department of Transportation. Prepared by the Highway Research Board and the Committee on Hazardous Materials, National Academy of Sciences, Washington, DC, 1969.
18. National Science Foundation. National Patterns of R&D Resource Funds and Manpower in the United States: 1953-1974. National Science Foundation, Washington, DC, 74-304, 1974.
19. National Transportation Safety Board. Noncompliance with Hazardous Materials Safety Regulations. National Transportation Safety Board, Washington, DC, Rept. NTSB-HZM-79-2, Aug. 3, 1979.
20. National Transportation Safety Board. On-Scene Coordination Among Agencies at Hazardous Materials Accidents. National Transportation Safety Board, Washington, DC, Rept. NTSB-HZM-79-3, Sept. 13, 1979.
21. National Transportation Safety Board. Railroad/Highway Grade Crossing Accidents Involving Trucks Transporting Bulk Hazardous Materials. National Transportation Safety Board, Washington, DC, Rept. NTSB-HZM-81-2, Sept. 24, 1981. NTIS: PB 82-113432.
22. National Transportation Safety Board. Standardization Maps of Hazardous Materials Accidents. National Transportation Safety Board, Washington, DC, Rept. NTSB-HZM-79-1, May 3, 1979.
23. National Transportation Safety Board. Federal Legislation Affecting the Transportation of Hazardous Materials with Selected Bibliographies and Data. Prepared for National Strategies Conference on Transportation of Hazardous Materials and Hazardous Wastes, Williamsburg, VA, Feb. 1981.
24. National Transportation Safety Board. Safety Effectiveness Evaluation--Federal and State Enforcement Efforts in Hazardous Materials Transportation by Truck. National Transportation Safety Board, Washington, DC, Rept. NTSB-SEE-81-2, Feb. 19, 1981. NTIS: PB 81-180101.
25. R. Nelson. World Leadership: The Technological Gap and National Science Policy. Minerva, Vol. 9, July 1971, pp. 386-399.
26. D.L. Price. Toward an Integrated Safety Training Program. Professional Safety, Vol. 21, No. 11, Nov. 1976, pp. 42-45.
27. D.L. Price and J.W. Schmidt. Voluntary Compliance: Effective Prevention in Hazardous Materials Transportation. Professional Safety, Vol. 24, No. 5, May 1979, pp. 18-21.
28. D.L. Price and J.W. Schmidt. A Breakthrough in Planning for Hazardous Materials Accidents. International Fire Chief, Vol. 46, No. 7, July 1980, pp. 17-19.

29. D.L. Price, J.W. Schmidt, and R.W. Kates. Multimodal Hazardous Materials Transportation in Virginia. Safety Projects Office, Virginia Polytechnic Institute and State Univ., Blacksburg, Rept. VDOTS/SPO-16, Sept. 1981.
30. Public Technology, Inc. Transportation of Hazardous Materials. Urban Consortium for Technology Initiatives/Public Technology, Inc., Washington, DC, Sept. 1980.
31. Rebecca L. Rawls. Chemical Transport--Coping With Disaster. Chemical and Engineering News, Nov. 24, 1980, pp. 20-30.
32. R.E. Rhoads. An Assessment of the Risk of Transportation of Gasoline by Truck. Pacific Northwest Laboratory, Richland, WA, Rept. PNL-2133 UC-71, Nov. 1978.
33. Sandia National Laboratories. Update (newsletter). Vol. 1, No. 3, Transportation Technical Center, Sandia National Laboratories, Albuquerque, NM, March 1982.
34. J.W. Schmidt and D.L. Price. Hazardous Materials Highway Transportation Accident Potentials in Virginia. Virginia Department of Transportation Safety; Transportation Safety Projects Office, Virginia Polytechnic Institute and State Univ., Blacksburg, Rept. VDOTS/SPO-9, Feb. 1980.
35. J.W. Schmidt and D.L. Price. The Flow of Hazardous Materials on Virginia Highways: 1978 Survey. Virginia Department of Transportation Safety; Transportation Safety Projects Office, Virginia Polytechnic Institute and State Univ., Blacksburg, Rept. VDOTS/SPO-10, April 1980.
36. J.W. Schmidt and D.L. Price. Flow of Hazardous Materials on Highways. Journal of Safety Research, Vol. 11, No. 3, Fall 1970, pp. 109-114.
37. Transportation Research Board. The Ten Most Critical Issues in Hazardous Materials Transportation. TRB, Transportation Research Circular 219, July 1980.
38. U.S. Congress, Senate, Committee on Commerce. Report on [the] Transportation Safety Act of 1974. Rept. 93-1192, 93rd Congress, 2nd Session, Sept. 30, 1974.
39. U.S. Department of Transportation. Transportation Safety Information Report--2nd Quarter, 1976. Transportation Systems Center, U.S. Department of Transportation, Cambridge, MA, 1976.
40. U.S. Department of Transportation. Tenth Annual Report: Hazardous Materials Transportation. U.S. Government Printing Office, 1980.
41. U.S. Department of Transportation. Eleventh Annual Report on Hazardous Material. U.S. Government Printing Office, 1981.
42. U.S. Department of Transportation. Twelfth Annual Report of the Administration of the Natural Gas Pipeline Safety Act. U.S. Government Printing Office, 1980.
43. U.S. General Accounting Office. Programs for Ensuring the Safe Transportation of Hazardous Materials Need Improvement. U.S. General Accounting Office, Washington, DC, Rept. CED-81-5, Nov. 4, 1980.
44. E. Wolf. Public Policy Toward Commercial By-Products from Government R&D. Presented at 49th Annual Conference of Western Economic Assn., Las Vegas, June 1974.
45. C.J. Wright. How Union Pacific Railroad Meets the Hazardous Material Challenge. Proc., Hazardous Material Spills Conference, Milwaukee, WI, April 19-22, 1982, pp. 271-274.
46. J.E. Zajic and W.A. Himmelman. Highly Hazardous Materials Spills and Emergency Planning. Marcel Dekker, Inc., New York, 1978.
47. J.C. Zercher. Emergency Reponse--A Changing Industry Attitude. Proc., Hazardous Material Spills Conference, Milwaukee, WI, April 19-22, 1982, pp. 231-237.

APPENDIX 2: Conference Attendees

William Andrews, PNL

Edward O. Baicy, Fragmentation Branch/BRL
Darrel J. Behrendsen, Colorado Training Institute
Tom Bell, Sajen Air, Inc.
Ludwig Benner, National Transportation Safety Board
Arthur Bensmiller, Transportation Safety Institute, Oklahoma City
Lawrence Bierlein, Attorney at Law
Mike T. Bohlman, Sea-Land Service, Inc.
Jerome Bone, New York State Department of Transportation
Don A. Boyd, DuPont deNemours and Company
William A. Brobst, The Transport Environment
Thomas W. Browne, UPS
Benjamin Buchbinder, Materials Transportation Bureau, U.S. Department of Transportation
Laura Buchbinder, FEMA, U.S. Fire Administration
Ronald Buckingham, Suffolk County, New York Department of Fire Safety
Noel C. Bufe, Northwestern University
William Buglass, Wisconsin Department of Transportation
William J. Burns, OST, U.S. Department of Transportation

Edmund J. Cantilli, Polytechnic Institute of New York
R.M. Cardillo, Exxon Chemical (Americas)
Lorenzo Casanova, Federal Highway Administration, U.S. Department of Transportation
Paul R. Chagnon, Military Traffic Management Command
Gregory Choppin, Florida State University
Joseph M. Clapp, Roadway Express
Albert C. Clark, Chemical Manufacturers Association
Adrian G. Clary, Transportation Research Board
Robert D. Coffee, Eastman Kodak Company
Charles R. Corbett, U.S. Coast Guard
George H. Cramer, Los Angeles Department of Transportation and Development

Russell Dawson, Business Publishers, Inc.
Ermes E. DeMaria, New England Nuclear Corporation
George B. Donaldson, Wilbur Ellis Company
Richard M. Doyle, American Trucking Associations, Inc.
John F. Funn, Illinois House of Representatives

James A. Echols, Sajen Air, Inc.
Robert D. Ervin, University of Michigan

Frank Farenchak, Pennsylvania Emergency Management Agency
G.G. Fleming, Chessie System
Conan P. Furber, Division of Environmental Studies, Association of American Railroads
John S. Gardinier, U.S. Coast Guard
Roy F. Garrison, U.S. Department of Energy
Dennis R. Garwood, Ohio Department of Highway Safety
Peter A. Gill, Family Lines Rail System
Theodore S. Glickman, Transportation Systems Center, U.S. Department of Transportation

Richard S. Golob, Hazardous Materials Intelligence Report
John Granito, State University of New York
Robert M. Graziano, OH Materials Company
Alfred W. Grella, Office of Inspection and Enforcement, U.S. Nuclear Regulatory Commission
Alexander J. Gretes, Maryland Institute for Emergency Medicine
M.B. Grove, Exxon Pipeline
Frank C. Gunderloy, Jr., Rockwell International
William H. Gushard, Greif Brothers, Springfield, New Jersey

Fred Halvorsen, U.S. Coast Guard
Bruce H. Hamill, National Paint and Coatings Association
Bob L. Hansen, Seattle Fire Department
J.Z.B. Harding, Consolidated Rail Corporation
Diane Harmon, Edlow International
William J. Harris, Jr., Association of American Railroads
Erskine E. Harton, Jr., Consultant
Edwin W. Hauser, Institute for Transportation Research and Education
Jerry A. Havens, University of Arkansas
Kevin Healy, New York City Department of Transportation
Mark F. Hengel, Missouri Pacific Railroad
David J. Hensing, American Association of State Highway and Transportation Officials
Morris Hershson, National Barrel and Drum Association
Arland V. Hicks, Kansas Department of Energy
Paul Hinchcliffe, Transportation Systems Center, U.S. Department of Transportation
Stanley Hoffman, Union Carbide

Jack Jacobsen, Profit by Air, Inc.
Robert M. Jefferson, Sandia National Laboratories
Will Jennings, Franklin, Tennessee
William E. Johns, American Trucking Associations, Inc.
H. Arvid Johnson, Container Corporation of America

William G. Kahler, Union Carbide Corporation
William Keffer, Emergency Planning Response, U.S. Environmental Protection Agency
James W. Kerr, Technological Hazards Research, FEMA
John Killilee, Consolidated Freightways
C. Charles Kimm, Battelle Columbus Laboratories
Paul V. King, Computer Sciences Corporation
Wendell M. Knight, Environmental and Safety Designs
Rolland G. Kuhlman, Burlington Northern Railroad
Myra Lee, Multnomah County, Oregon, Emergency Management
Donald Levine, Rail Vehicle Safety Research Division, Federal Railroad Administration, U.S. Department of Transportation
J.R. Ligon, Consultant

Philip P. Madonia, Illinois Department of Transportation

John W. Marshall, Edwards Air Force Base (Rocket Propulsion Laboratory)
Steven McDougall, Industrial Hygienist, Teamsters Union
Paula H. McGrew, Price, Williams and Association
Rosemary Menard, Seattle City Council
Robert J. Mesler, Jr., Dow Chemicals (U.S.A.)
Stephan Michelson, Econometric Research, Inc.
Arnold Moodie, Fluor Engineers and Contractors

Hyla Napadensky, IIT Research Institute
John C. Nordin, Illinois State Police
Donald Nussbaumer, U.S. Nuclear Regulatory Commission

J.J. O'Driscoll, Southern Railway System
Emmett O'Hare, Consultant
Floyd E. Ouellette, Dow Chemicals (U.S.A.)

John A. Pachuta, Pennsylvania Department of Transportation
Mike Parnarouskis, U.S. Coast Guard
Robert L. Paullin, Materials Transportation Bureau, U.S. Department of Transportation
Thomas A. Phemister, Association of American Railroads
Lloyd L. Philipson, J.H. Wiggins Company
Kenneth L. Pierson, BMCS, Federal Highway Administration, U.S. Department of Transportation
Suellen Pirages, OTA, U.S. Congress
Thomas W. Powers, Maryland Department of Transportation
Simon Prensky, Transportation Systems Center, U.S. Department of Transportation
Dennis L. Price, Virginia Polytechnic Institute and State University

Elaine Randolph, Los Angeles Times
Robert G. Rhodes, A.T. Kearney, Inc.
Robert W. Riedl, Milwaukee, St. Paul and Pacific Railroad Company
Alan I. Roberts, Materials Transportation Bureau, U.S. Department of Transportation
Frank Roberts, Cedar Falls Corporation
Robert Robison, Portland Office of Emergency Services
William D. Rowe, American University
Deborah Rudolph, Office of the Secretary, U.S. Department of Transportation
Eugene R. Russell, Kansas State University
Harvey Ryland, ISE TAP/OSTP

Lee D. Santman, Materials Transportation Bureau, U.S. Department of Transportation

Raymond D. Scanlon, Port Authority of New York and New Jersey

Amy Schaeffer, Office of Enforcement, U.S. Environmental Protection Agency
William R. Schaeffer, Monsanto Chemicals
Lawrence Scheufele, Illinois State Police
J.W. Schmidt, Virginia Polytechnic Institute and State University
W.R. Seebaugh, Fried, Frank Harris, Shriver and Kampleman, Attorneys
Dick Sexton, Hazardous Materials Transportation Newsletter
H.M. Shappell, Shipping Container Institute
Ronald M. Sharp, Norfolk and Western Railway Company
Donald Shilesky, SCS Engineers
Pete Sill, RSPA, U.S. Department of Transportation
Ben Sosewitz, Envirodyne Engineers
Bonnie E. Southwick, Tri-State Motor Transit Company
Marjorie Stailkovich, Transport Canada, Ottawa
Gustav H. Strobel, Travelers Insurance Company
Anne Stubbs, Council of State Governments

David Teeter, Puget Sound Council of Governments
C.H. Thompson, Aerojet General Corporation
David Thompson, Thompson Tank and Manufacturing, Inc.
Kenneth A. Thompson, Yellow Freight System, Inc.
Edgar L. Tidd, Virginia Department of Highways and Transportation
Leonard M. Trosten, LeBoeuf, Lamb, Leiby and MacRae
Paul R. Tutt, University of Tennessee

Gary L. Urbanek, Allard, Inc.

Jack Vallas, California Highway Patrol
Karsten J. Vieg, Illinois Department of Transportation
Gordon O. Voss, Minnesota House of Representatives
Stanley N. Wasser, Illinois Department of Transportation
Marge Weissman, U.S. Department of Transportation, Seattle
L.R. Welch, Texas Eastern Transmission Products Pipeline
Hillary Whitaker, National Governors Associations
A.D. Williams, Union Pacific Company
Mary M. Williams, U.S. Coast Guard
Robert C. Williams, National Highway Traffic Safety Administration
Donna Wise, National Conference of State Legislators
David Woodbury, Wisconsin Emergency Government
Jane Woodward, Performance Development Institute
C.L. Wright, Jr., Pennsylvania Turnpike Commission
Charlie J. Wright, Union Pacific Railroad
James V. Zaccor, Scientific Services, Inc.
John C. Zercher, Chemical Manufacturers Association
Robert C. Ziegler, Chemical Lehman Tank Lines
John Zogby, Pennsylvania Department of Transportation

APPENDIX 3: Biographical Data on Steering Committee to Develop a National Strategy for the Transportation of Hazardous Wastes in the 1980s

SCANLON, RAYMOND D., (Chairman, Steering Committee, October 1982-) Administrator of safety program, hazardous materials specialist, Port Authority of New York and New Jersey; born U.S.A., March 1937. B.A. Economics, Fordham University; Master of Public Administration, Fairleigh Dickinson University; responsible for safety standards for Port Authority Facilities, conferring with patrons, government officials, shippers, manufacturers, etc., to explain and assure compliance with federal, state, and local regulations pertaining to hazardous materials, hazardous wastes, and toxic substances; training policy and civilian personnel within the Port Authority, consulting for Port Authority Aviation and Port Departments, and assisting the Authority's Law Department and Police Division on litigation proceedings. Chairman of U.S. DOT Hazardous Materials Transportation Committee, Region II; Consultant to National Fire Protection Association; Advisory Board Member, Puget Sound Council of Governments; Member, Mayor's Task Force (NYC) on Hazardous Materials Transportation.

VIEG, KARSTEN J., (Chairman, Steering Committee, 1980-1982) Traffic Safety Administrator, state highway transportation official; born U.S.A., December 12, 1934; Antioch College, B.A.; University of California, J.D.; Illinois Governor's Representative for Highway Safety and Director, Division of Traffic Safety, Illinois Department of Transportation, since 1974; Staff member, State of California and Midwest Research Institute prior to 1974. Currently heads a staff of 260 with an annual budget of approximately $20 million. The Division of Traffic Safety has jurisdiction over administration of Section 402 highway safety funds and the regulation of the Illinois Hazardous Materials Transportation Regulations over highways. In addition, the Division is responsible for Illinois' vehicle inspection programs and is the office of record for all highway accident data.

BIERLEIN, LAWRENCE W., Lawyer, hazardous materials specialist; born Cleveland, Ohio, May 4, 1942; B.A. Johns Hopkins 1964, J.D. University of Pennsylvania 1967. Private practice since 1972. Prior to that time served in Office of the General Counsel of U.S. DOT. Author of the Red Book on Transportation of Hazardous Materials, 1977. Currently Editorial Director of a monthly newsletter, "Hazardous Materials Transportation."

BROBST, WILLIAM A., Consultant, chemist; U.S. citizen born in 1930. B.S. in Chemistry from Northwestern University 1951; President, The Transport Environment, consultant in hazardous materials transportation since 1979; Chief, transportation safety programs for DOE (and its predecessors ERDA and AEC), manager of DOE's transportation systems research, development, and testing programs for a variety of fuels and wastes 1970-1979; Deputy Director, DOT's Office of Hazardous Materials 1966-1970; previously served in a similar position with the ICC; head of the radiological physics program in AEC's Chicago office; naval officer with extensive nuclear weapons storage and testing experience. Chairman, Transport Advisory Group, International Atomic Energy Agency.

CHOPPIN, GREGORY R., Educator, chemist; born Eagle Lake, Tex., Nov. 9, 1927. Education: Loyola Univ. La., B.S. 1949; Univ. Tex., Ph.D. (Chem), 1953; Loyola Univ., DSc, 1969. Prof. Exp: Mem. staff, Radiation Lab., Univ. Calif., 1953-1956; from Asst. Prof. to Assoc. Prof., 1956-1963, Chmn dept., 1968-1977, Prof. Chem., Florida State Univ. 1963-present. Concurrent Pos: Vis Scientist, Ctr Study Nuclear Energy, Belgium, 1962-1963; Fulbright Lectr., Uruguay, 1965, and Portugal, 1969; Vis. Prof., Sci Univ. Tokyo, 1978. Mem: AAAS; Am Chem Soc (chmn, Div. Nuclear Chem Technol, 1976). Res: Nuclear chemistry, physical chemistry of the actinides and lanthanides; structure of water and aqueous solutions. Chmn, Subcommittee on Nuclear and Radiochemistry of the Committee on Chemical Science, National Research Council.

GRAZIANO, ROBERT M., Corporate executive, hazardous materials specialist; U.S. citizen born May 26, 1940. Villanova University, business and social sciences (1958-1960); currently vice president, government and industry relations, OH Materials, Inc.; 1970-1979, Director, Bureau of Explosives, Association of American Railroads; 1969-1970, responsible for coordination and development of Bureau's tank car program; 1962-1968, various positions with Long Island Railroad and AAR.

HARTON, ERSKINE E., JR., Consultant (hazardous materials and system safety), Chemist; U.S. citizen born August 1919. B.S. in Chemistry, Geneva College; M.S. in Chemistry, University of Wyoming; 1979-present, Consultant; 1979, Acting

Director, Office of Program Support, Materials Transportation Bureau (MTB), U.S. DOT; 1978-1979, Chief, R&D Management Div., MTB, U.S. DOT; 1969-1979, Chemical Engineer, Technology Div., Office of Hazardous Materials Operations, U.S. DOT; 1968-1969, Chief, Hazardous Cargos Branch, National Highway Safety Bureau, U.S. DOT; 1968, Functional Manager for Hazards Identification, Safety Research and Data, National Aeronautics and Space Administration (NASA); 1967-1968, Acting Safety Director, NASA; 1961-1968, Assistant Safety Director, NASA: 1960-1962, Chemical Engineer, 2705th Air Munitions Wing, U.S. Air Force; 1955-1960, Associate Chemist, Midwest Research Institute; 1948-1955, Chemical Engineer, U.S. Bureau of Mines Petroleum & Oil Shale Expt. Station; 1942-1948, Chemist, Pure Oil Research Laboratory. Fellow, American Institute of Chemists; Past President, Washington, D.C. Chapter, System Safety Society; member, American Chemical Society, American Society of Safety Engineers, Sigma Xi-RESA and the North American Thermal Analysis Society. Certified Professional Chemist; Certified Safety Professional, Registered Professional Engineer (Illinois); Registered Professional Safety Engineer (California).

HAVENS, JERRY A., Educator; Chemical Engineer; born November 24, 1939; U.S. citizen. B.S. Chem. Engrg., University of Arkansas, 1961; M.S. Chem. Engrg., University of Colorado, 1962; Ph.D. Chem. Engrg., University of Oklahoma, 1969; Post-Doctoral Studies, Flame Dynamics Laboratory, University of Oklahoma, 1969. Instructor, Chem. Engrg., University of Oklahoma, 1967-1968; Assistant Professor, Chem. Engrg., University of Arkansas, 1970-1974; Associate Professor, Chem. Engrg., University of Arkansas, 1974-1979; Professor, Chem. Engrg., University of Arkansas, 1979-present. Member, Sigma Xi Research Fraternity; American Chemical Society; Tau Beta Pi; American Inst. of Chemical Engineers; American Assn. for the Advancement of Engrg. Education; National Fire Protection Assn.; American Society for Testing & Materials - Past Chm., Subcommittee on Thermal Stability of Condensed Phase Chemicals and Past Vice-Chm., Committee on Hazard Potential of Chemicals. SOHIO Fellowship, University of Colorado, 1962; Union Carbide Fellowship, Univ. of Oklahoma, 1966-1967; Registered Professional Engineer, Arkansas.

JEFFERSON, ROBERT M., Transportation Technology Development, federal official; Mechanical engineer; born Akron, Ohio, February 10, 1932; BSME Michigan Tech. University 1953; MBA, University of New Mexico 1964. Sandia Laboratories (24 years) and, currently, Manager, Nuclear Materials Transportation Technology Department; Transportation Technology Center and University of New Mexico (17 years) as Adjunct Professor of Nuclear Engineering. Member New Mexico Energy Institute, American Nuclear Society and of the Society's Publications Steering Committee and Admissions Committee. Listed in Who's Who in Atoms, Who's Who in the West, and Who's Who in Public Service. Past member, NAS Committee on Nuclear and Alternative Energy Sources; Chairman, 5th International Symposium on Packaging and Transportation.

JOHNS, WILLIAM E., Trucking Association Executive. Born U.S.A., January 6, 1927; Graduate, University of Maryland, B.S., transportation; 1953 to present, Managing Director of the Technical Services Division of the American Trucking Associations, Inc. (ATA), Director of ATA's Dept. of Safety and Security; before 1953 various assignments for Associated Transport, Inc. Member of the Board of Directors of the National Safety Council and of the NSC Motor Transportation Conference Executive Committee. Past Chairman of the Hazardous Materials Advisory Committee of the Transportation Association of America and member, Executive Committee of the National Committee for Motor Fleet Supervisor Training.

LAMKIN, JACK T., Educator, Economist; born, Brownswood, Texas, 1935; B.S. Agricultural Economics, Texas A&M, 1957; M.S. Economics, Texas A&M, 1967. Program Manager, Texas Transportation Institute, 1978-present, Research economist, Rail systems program, transport operations, transport economics 1965-1977; Secy-Treas. Lamkin Bros., 1959-1965; Consultant. Member, Transportation Research Forum, Omicron Delta Epsilon, Delta Nu Alpha.

O'DRISCOLL, JEREMIAH J., Railroad official; safety engineer; born Galveston, Texas, Dec. 8, 1925; Graduate Navy V-12 Program, 1944, Southwestern La. Institute; BBA, University of Texas, 1950; Director of Hazardous Materials and Safety, Southern Railway, 1969-present; Mgr. Corp. Safety and Loss Prevention, Atlas Chemical Inc. 1959-1969; 1943-1947 and 1950-1954, U.S. Navy, Explosive Safety Engineer, Atomic Defense Officer. Certified safety professional, Prof. member--American Society of Safety Engineers, Prof. Engr-California, Who's Who in Finance and Industry. Publications include "Permit Systems: in "Safety and Accident Prevention in Chemical Operations "(J. Wiley & Sons); "Emergency Action Plan for Hazardous Materials Incidents" (Southern Railway System 1969), "Transportation Emergency Action Guide for Hazardous Materials Incidents" (Jody, Inc., 1977). Member, Committee on Hazardous Materials, NAS-NRC.

PRICE, DENNIS L., Educator, industrial engineer; born U.S.A. October 23, 1930; Ph.D., Industrial Engineering, Texas A&M, 1974; currently Associate Prof., Dept. of Industrial Engineering and Operations Research, and Director, Transportation Safety Projects Office, Virginia Polytechnic Institute and State University; formerly employed by the Autonetic Division of North American Rockwell Corp.

RUDOLPH, DEBORAH KAREN SCHULTZ, Federal transportation agency staff director; born U.S.A. June 28, 1947; B.S. American University 1974, M. Bus. Administration Studies, Arizona State Univ. 1977-78. Currently Staff Director for the Transportation, Commerce & Community Development Task Force of the White House Office of Science and Technology Policy's (OSTP) Intergovernmental Science, Engineering and Technology Advisory Panel; Phoenix, Arizona, Management and Budget Dept. 1975-1978; White House OMB, Congressional Relations Office 1974.

RUSSELL, EUGENE R., Educator, civil engineer, born Connecticut, August 24, 1932; B.S. CE. Univ. of Missouri 1958; M.S. Iowa State Univ. 1965; Ph.D. Purdue Univ. 1974. Assoc. Prof. of CE Kansas State Univ. 1974 to present; Graduate Instructor, Purdue 1972-73; Research Engineer, Purdue's Highway Extension and Research Project for Indiana Counties 1969-1971. Asst Prof. CE,

Indiana Institute of Technology, 1965-1968. Asst. Bridge Engineer, Calif. Division of Highways, 1958-1962. Part-time soils consultant for L.G. Petro and Associates 1965-1969. Author of numerous papers and reports on soils, traffic engineering, and, recently, research programs to develop community models for handling hazardous materials transportation emergencies. Donald T. Davidson Memorial Award 1963-64. Member Chi Epsilon, Sigma Chi, ASCE, ASEE, NACE, ARSA.

WILSON, GEORGE L., JR., (deceased) Association Executive, Economist; Born Philadelphia, PA, May 25, 1919. B.Sc. in Economics (Cum Laude), Wharton School, Univ. of PA. 1940 (Transportation); MBA 1948, Temple Univ. (Transportation Economics), Univ of PA Graduate School 1948-1949, Industrial Management. Part-time educator 1947-1957 "Principles of Economics and Transportation Economics" Temple Univ. and St. Joseph's College, Philadelphia. 1976-present, Managing Director, Hazardous Materials Advisory Council; 1973-1976, Director of Traffic, North American Region, Rohn and Haas Co.; 1948-1955, various assignments with Rohn and Haas Co.; 1940-1942, 1946-1948, Traffic Dept. and Traffic Coordination Dept., Atlantic Refining Co. Associations; 1969-1971, Chairman, Trans. and Distribution Committee, Manufacturing Chemists Association (MCA); 1961-1965, Chairman, Tank Car Compensation Committee, MCA; admitted as practitioner, ICC, 1948; Beta Gamma Sigma; 1961-1969, Vice President and Director, General Alumni Society, Univ. of Pennsylvania.